C.R. Tumblin
CONSTRUCTION COST ESTIMATES

Harvey V. Debo and Leo Diamant
CONSTRUCTION SUPERINTENDENTS JOB GUIDE

Oktay Ural, Editor
CONSTRUCTION OF LOWER-COST HOUSING

Robert M. Koerner and Joseph P. Welsh
CONSTRUCTION AND GEOTECHNICAL ENGINEERING
USING SYNTHETIC FABRICS

J. Patrick Powers
CONSTRUCTION DEWATERING: A GUIDE TO THEORY
AND PRACTICE

Harold J. Rosen
CONSTRUCTION SPECIFICATIONS WRITING:
PRINCIPLES AND PROCEDURES, Second Edition

Walter Podolny, Jr., and Jean M. Mu^{3}ller
CONSTRUCTION AND DESIGN OF PRESTRESSED
CONCRETE SEGMENTAL BRIDGES

Ben C. Gerwick, Jr. and John C. Woolery
CONSTRUCTION AND ENGINEERING MARKETING
FOR MAJOR PROJECT SERVICES

James E. Clyde
CONSTRUCTION INSPECTION: A FIELD GUIDE TO PRACTICE,
Second Edition

Julian R. Panek and John Philip Cook
CONSTRUCTION SEALANTS AND ADHESIVES, Second Edition

Courtland A. Collier and Don A. Halperin
CONSTRUCTION FUNDING: WHERE THE MONEY COMES
FROM, Second Edition

James Fullman
CONSTRUCTION SAFETY PRACTICE

Construction Funding

Construction Funding

Where the Money Comes From

Second Edition

Courtland A. Collier, P.E.

Associate Professor, Civil Engineering
University of Florida

Don A. Halperin, F.A.I.G.

Director Emeritus, School of Building Construction
University of Florida

A Wiley-Interscience Publication

John Wiley & Sons

New York • Chichester • Brisbane • Toronto • Singapore

Library of Congress Cataloging in Publication Data:

Collier, Courtland A.
  Construction funding.

  (Wiley series of practical construction guides)
  Rev. ed. of: Construction funding / Don A. Halperin.
1974.
  "A Wiley-Interscience publication."
  Includes index.
  1. Construction industry—Finance. 2. Mortgage loans.
3. Leases. I. Halperin, Don A. II. Halperin, Don A.
Construction funding. III. Title. IV. Series.
HD9715.A2C59 1984   690'.068'1   83-21753
ISBN 0-471-89065-0

Printed in the United States of America

10 9 8 7 6 5 4 3 2 1

Series Preface

The construction industry in the United States and other advanced nations continues to grow at a phenomenal rate. In the United States alone construction in the near future will approach two hundred billion dollars a year. With the population explosion and continued demand for new building of all kinds, the need will be for more professional practitioners.

In the past, before science and technology seriously affected the concepts, approaches, methods, and financing of structures, most practitioners developed their know-how by direct experience in the field. Now that the construction industry has become more complex there is a clear need for a more professional approach to new tools for learning and practice.

This series is intended to provide the construction practitioner with up-to-date guides which cover theory, design, and practice to help him approach his problems with more confidence. These books should be useful to all people working in construction: engineers, architects, specification experts, materials and equipment manufacturers, project superintendents, and all who contribute to the construction or engineering firm's success.

Although these books will offer a fuller explanation of the practical problems which face the construction industry, they will also serve the professional educator and student.

M. D. MORRIS, P.E.

Preface

The purpose of this book is to provide builders and developers with some of the vital tools needed to transform a project dream into a brick-and-mortar reality. The success of most projects is hinged upon proper financing, so a thorough knowledge of how to obtain good financing is a necessary asset for anyone moving upward on the ladder of success in the construction industry. This book provides a step-by-step guide through each important phase of financing a project, whether it be a simple one-family residence or a large multiunit complex. It is well recognized that the rewards in this industry are directly proportional to the skill and ability of the individual. This text should greatly improve those skills and the rewards that follow should grow correspondingly. The mastery of the knowledge contained in this text should prove a satisfying and rewarding experience in many ways to both the experienced practitioner and the ambitious beginner.

We would like to express our gratitude to the many enterprising financial advisors who have worked together with their construction industry counterparts blazing new trails, introducing what seemed at the time like radical new financial concepts but now have become routine methods for financing construction projects. The contributions made by these bold innovators provide the foundation on which we now proceed to even newer innovations in financing.

We also wish to gratefully acknowledge the patience, encouragement, and support from our wives, families, and secretaries, who have endured with us the long hours of travail that is a necessary prelude to attaining the ultimate satisfaction of producing a much needed and practical text that this book represents. We are pleased to dedicate this work to them for without their contributions, this work would have come to fruition. Jeff Vaughn also deserves commendation for his contribution.

COURTLAND A. COLLIER
DON A. HALPERIN

Gainesville, Florida
December 1983

Contents

Appendixes

1

How to Get Started as a Successful Builder/Developer

1

Introduction: Development Projects from Conception to Demolition

1.1. Introduction

It all begins in the mind's eye. Amid the inner recesses of the creative mind of a venturesome developer, a few wisps of an idea begin to come together. The idea takes shape first as a promising concept, then, gathering substance, it grows and finally springs forth as a full blown plan for a development project. In the marketplace, people with money have an unmet need. This newly conceived project will meet that need. The people in the marketplace will spend their money to have their needs satisfied. And in this way another worthwhile construction project begins its eventful journey through life.

Looking across the broad horizon of time that encompasses the lifetime of any man-made project, the project's conception is just the beginning of a series of phases that all projects flow through, from project conception in the mind's eye, on finally to demolition, hopefully at the end of a long and productive life. Sometime, somewhere, the project begins life as a mental image in someone's mind. It grows from there. Passing through increasingly rigorous feasibility tests, it takes on more distinct form and substance while moving through the design and construction phases, and on to occupancy and operational use. Finally in the latter days, perhaps after undergoing a renovation or two, the project is considered for demolition and replacement and finally is terminated as a newer and more productive project is conceived to replace it. Each of these project stages is associated with a different type of cash flow and need for financing. The overall view of the complete cycle is presented here in order to put the construction funding portion of that cycle into proper perspective.

1.2.   Conception

Before reaching the design and construction stages, all projects had to begin as an idea in someone's mind. In the private enterprise sector of the economy new ideas for construction are often generated as a result of someone's perception of some yet unfulfilled need in the public marketplace. For instance, someone may note a shortage of housing or office space, or the desire by industry to locate or expand in some area which lacks an adequate selection of industrial parkland. Realizing that one of the surest ways to earn a profit is to fulfill some public market need with a quality product reasonably priced, someone will begin serious consideration of a suitable development project to fulfill those needs.

In the public (government) sector of the economy, the needs of the tax-paying public are constantly monitored by responsible public officials as well as by user groups. Does traffic congestion indicate a need for new road projects, better traffic signalizations, or more mass transit facilities? Does community growth require extension of utility lines, expansion of utility plants or public service facilities such as police and fire stations? Does more leisure time indicate a need for a new library, more recreation facilities, concert halls, or convention centers? Someone, somewhere, recognizes a need of the taxpaying or ratepaying public and conceives an idea for constructing a facility to fulfill that need.

1.3.   Feasibility

While all projects begin as ideas, not all ideas become projects. The key question at this stage is "Will the idea work?" Only a small percentage of ideas withstand the acid test of feasibility and proceed on toward design and construction. Preliminary feasibility studies usually begin quite informally on a rule-of-thumb basis. If the residential rental market will sustain a monthly rent rate of 0.015 times the construction cost, the concept for a proposed residential rental project may have some promise and is moved forward for more detailed study. If land for residential construction is available at a cost not exceeding 20% of the development and construction costs, then land acquisition holds promise of passing the feasibility tests. Is money available to finance the project? Can we find a competent, reliable, and trustworthy contractor who can complete the job on schedule? If these informal preliminary feasibility studies produce promising results, then the formal feasibility studies can begin. Very little expenditure of time and money have occurred up until now, but formal feasibility studies usually require a significant input of both. Formal feasibility studies involve the gathering and analysis of actual data on a real market. What is the existing market supply, what projects does the competition have already underway or in the planning or proposal stages? What is the current and projected demand for the product? Is money for financing available at a reasonable cost? Taking into account all of the costs and incomes involved, can the product be marketed at a competitive price and still make an attractive profit? Is there a conservative margin for error or is this a high-risk speculative undertaking? Feasibility studies can be conducted by anyone knowledgeable in the markets involved, including the investor, the developer, or an outside firm specializing in feasibility studies. The feasibility study is frequently summarized into a written brochure for later use as part of the financing presentation designed to persuade lenders or investors to provide financing for the project.

1.4. Land Acquisition and Tentative Financing

After the feasibility studies indicate that there is a reasonable probability of success for the proposed project, two important steps must be taken: (1) the land on which to construct the project must be secured, and (2) a commitment by some lender to finance the project must be obtained. One popular method used by conservative developers to secure the land involves obtaining an option to purchase the land. An option on the land is the right for a limited period of time to purchase the land, and the option sometimes can be obtained at a very low cost as will be explained in Chapter 5. This option minimizes the loss of money if for any reason construction does not go forward.

Other methods of securing the land include outright purchase, long-term lease, offer of a minority partnership share or other interest to the land owner, and so on.

Availability of financing must be determined at this point also. A builder will find it essential to develop a general familiarity with the financial markets that have capabilities of funding such projects. In addition a builder must also develop access to key contact people within those markets (such as bank loan officers or mortgage brokers). One or more possible sources of funding for the project should be located by this point in time. Of course, at this stage, potential lenders usually will qualify their preliminary financial commitments with reasonable performance requirements by the borrower. The lenders might agree to make the loan *providing* that the borrower obtains a satisfactory design, a reasonable, low bid price from contractors, and that interest rates plus points (a fee charged for obtaining the loan) and other costs are within the range contemplated during the feasibility study.

1.5. Design

As the original idea was being formulated during the conception stage, its form and substance may have been somewhat fuzzy and indistinct, and perhaps it actually consisted of a range of similar alternative proposals. During the feasibility stage, the project outline proceeds to take on more definite shape. By the time the design stage is reached the project by necessity has arrived at clear-cut fruition. The owner confers with a designer, usually a professional engineer or architect, and describes all the functions the project is supposed to perform as well as its budget limitations and general appearance. The designer then translates those thoughts into design concepts and then into plans, specifications, and contract documents. These design services are typically provided by professionals under contract at a cost between about 6 and 10% of the total project cost, depending on the complexity of the design.

1.6. Construction

After the design is completed, the financing committed, the land secured, the permits obtained, and decisions reached on any other consequential details, the contract is awarded for construction of the project. The form of construction contract may be any one of a standard variety of forms or some progressive modification thereof. The construction contract may be awarded as a result of competitive bidding or negotiations or both, as discussed in later chapters. The construction period may consume from several months to several years depending on the size and complexity of the project. Financially the construction period can be a trying time for the owner for two reasons: (1) the cash flow is usually all negative (the owner has a lot of bills to pay, thus the cash flows *out* of the owner's accounts, with no income from the project while it's still under construction), and (2) the cash flow is often a larger negative amount than anticipated due to unforeseen contingencies and last-minute changes (as the project begins to take shape during construction, many owners are stimulated to think of changes that improve the project but also raise the cost). In addition, if completion is late, the repayment of the construction loan must be delayed with corresponding increases in the interest charges to cover the construction loan. Similarly the contractor building the project must plan his cash flow very carefully to cover payroll, materials purchases, and equipment payments. Many of the contractor's bills must be paid several weeks or even months before receiving reimbursement from the owner, thus leading to a frequent need to borrow over a short term to pay obligations incurred in anticipation of payment from the owner.

1.7. Operation over a Normal Life

After the project is completed, the short-term construction loan is repaid with funds borrowed under long-term mortgage financing. The intended users then move into the project and operate and maintain it over a normal life span. For buildings, the life span usually ranges from 40 to 100 years, with a few lying outside these ranges. The life of most buildings end when they become obsolete, which usually occurs before they wear out or fall into disrepair and decay due to lack of maintenance. A well-maintained building will last for hundreds, if not thousands, of years. Many buildings acquire a subsequent "new lease on life" by means of remodeling or renovation.

1.8. Demolition and Replacement

From as early as the time of construction completion, a project begins competing with other alternative project concepts for the use of the land it occupies. Would a newer, larger structure on that site be more profitable or serve a higher need? The existing project will be re-evaluated frequently during its lifetime and compared to other possible projects competing for that location. As the existing project ages, it usually becomes more vulnerable to displacement by a newer, more productive project concept. And, ultimately, a new and potentially better concept finally does become successful in challenging and displacing the existing project. The old structure is demolished, a new one begins to take shape, and the cycle begins again.

2

Construction and Development Firms: Four Ways to Set Up a Successful Business, and How to Raise Capital for Each

2.1. Goals and Needs for Capital

The goal of business is to earn a good return on its investment of money, time, and other resources by providing outstanding products or services at competitive prices. In order to achieve that goal, the business needs (1) capital, and (2) an efficient form of company organization.

2.2. Capital—Long-Term Fixed Assets

To successfully operate a business even the smallest company must have a certain amount of capital. This capital usually takes two forms: (1) Long-term "fixed" assets, and (2) Short-term or "liquid" assets. Long-term assets consist of property that has value, but the value is not quickly nor easily turned into cash. For example, land, furniture, used construction equipment, and vested interest in a retirement fund are all long-term assets. All of these have value and would be listed on a financial statement as assets, but a lender would not count them as valuable as cash for purposes of securing a loan. The lender is viewing the borrower's assets as sources of funds to repay the loan if all else fails. To turn someone else's land or furniture into cash is a long and tedious process, therefore those types of long-term assets are *not* highly regarded by lenders. Some lenders will discount such long-term assets by as much as 50% of market value. Thus

$10,000 cash is counted as a $10,000 liquid asset, but the same $10,000 invested in land or equipment may be counted as only $5000 worth of asset value for purposes of securing a loan.

2.3. Capital—Short-Term Liquid Assets

The second type of capital is referred to as short-term or "liquid" assets. Cash in the bank is an example of a *liquid asset*, because it can be converted into cash almost immediately. Other typical examples of liquid assets include stocks, bonds, and notes receivable, providing the issuers have good credit. Banks and other lenders are naturally more favorably disposed toward lending money to companies with sizable liquid assets. In fact, they frequently will lend several times the actual value of those assets. Therefore liquid assets allow for leverage, because a smaller amount of assets can move a much larger loan amount. Financial leveraging is somewhat analogous to the physical leveraging of Archimedes, who claimed he could move the world if he had the proper lever and fulcrum. With a relatively small amount of liquid assets, a large amount of property can be controlled through proper leveraging.

Thus, liquid assets are necessary to obtain loans that in turn provide the cash needed by construction firms to finance their construction contracts. Generally a construction firm should limit its contract volume to no more than about 10 times its liquid assets. For example, a firm with liquid assets of $100,000 normally should *not* take on contracts totalling more than $1,000,000 at any one time. The company's liquid assets should be carefully guarded, and major expenditures of cash, although sometimes necessary, should be undertaken with great caution. One of the major hurdles in starting up a construction firm is to develop an adequate amount of operating capital to start with and then add to that sum for growth as time goes on.

2.4. Forms of Company Organizations

There are four standard forms of company organization under which a construction firm can be organized.

1. Individual owner (sole proprietorship)
2. Partnership (any number of partners)
3. Limited partnership (general partner and any number of limited partners)
4. Corporation (stockholders, board of directors, president, and other employees)

Each of these forms of business has certain advantages and disadvantages with regard to authority and responsibility of the principals (owners and operators). In addition, each form of business has an effect on the ability to raise funds needed for the operation and growth of the company. The methods of borrowing under each form are similar. Wherever the collateral is marginal, the lender may require that some individual with good credit (such as the president of the company, or the partner with the most liquid assets) will agree to be personally responsible for any loan notes. For a construction loan this guarantee might extend up until the time construction is completed and the permanent mortgage loan is in force. After that time the mortgage lender has a finished property as security, and usually no longer requires an individual to assume personal liability.

2.5. Individual Proprietorship

Undoubtedly the construction and development industry offers some of the greatest opportunities for those who aim to make a fortune and don't mind hard work. The starting capital available to many now-successful contractors was surprisingly small when they first started out on their own. Today those who want to start contracting on their own can still accumulate capital in the old-fashioned way, combining hard work, morality, and self-denial. The hard work produces cash, the morality produces the good reputation for honesty and integrity among lenders and customers, and self-denial creates the ability to reduce needless spending, hold on to the cash, and accumulate more savings. Two other alternatives are to have rich parents or to marry a wealthy spouse, but these courses are not commonly available. Instead, the regular income produced by hard work, coupled with a steady savings plan can be relied upon to yield the necessary dollars in a reasonable amount of time.

2.6. Getting Started

Starting with about $10,000, a sole proprietor builder may expect to erect one house worth about $100,000, or two $50,000 houses. The builder should carefully select the right location, choose the type of house that appeals to a popular market, and give a good value for a fair price. Given average luck, the houses should sell in less than 4 months after construction starts, with a net profit of perhaps 5%, or $5000. Net profit should be calculated as the remaining income left after all costs have been paid, including the buider's wages. Remember that the boss is working for the company too, and is therefore entitled to a reasonable salary. The salary is part of the company's cost of doing business and should be accounted for in job overhead charges. Notice that the profit of $5000 is only 5% of the volume of business that was accomplished, but amounts to an increase of 50% of the initial capital.

2.7. Growing with Liquid Assets

If the houses can indeed be sold within 4 months after starting, then in 1 year's time the fledgling builder potentially can turn over the original capital funds three times and make a total profit of $15,000 in addition to wages. If this profit is added to the original capital, the total capitalization might now stand at $10,000 + $15,000 = $25,000. With liquid assets of $25,000, the company can increase the volume of business to 10 × $25,000 = $250,000 in each 1-year business cycle. The anticipated profit will now be 0.05 × $250,000 = $12,500 at the end of each 4-month period, leading to a gain of $37,500 for the second year. Now the bank account stands at $25,000 + $37,500 = $62,500, in addition to any salary taken from the business. And with capitalization of $62,500 the builder is ready to start on some larger projects.

2.8. Fringe Benefits—Tangible and Intangible

One of the richest rewards of a sole proprietorship is the feeling of pride and accomplishment that comes when the company grows to the level of a successful ongoing operation. Achievement of this status depends upon the expertise and decisions of the company's founder, and success for the company means success for the individual. No reward can be richer. Additional fringe benefits to the sole proprietor include the rightful ability to exercise the company funds for the financial benefit of the owner. For example, the company should pay for some life insurance on its owner. The owner is the key person, and without him/her the business might fail. In the event of the owner's death the insurance could be used to help pay off current debts and place the beneficiaries in a much better position to carry on the business, should they so desire. Furthermore, the company can often employ members of the owner's family resulting not only in extra work and extra pay but also some tax advantages on income sharing. Remember that the Internal Revenue Service insists that any employment of family members resulting in a tax advantage be genuine and not fictitious. A competent accountant can list other advantages of individual ownership.

2.9. Liability of a Sole Proprietorship

Along with the responsibility of individual ownership there also comes liability. If instead of making money the company should lose money, the losses normally can reach beyond the company's assets into the owner's personal assets. Any borrowing done by the company actually becomes personal borrowing, constituting a personal indebtedness, even though a company structure exists. For example, if Raymond Cornpone is doing business as Cornpone Building Company, and he signs all notes R. Cornpone, d.b.a (Doing Business As) Cornpone Company, he will still be personally liable for the notes. If the Cornpone Company is forced into bankruptcy, he himself will have to make good on the notes or else undergo personal bankruptcy, which does not always get rid of all debts. Furthermore, his personal liability usually extends to all company operations. If a child is severely injured while playing on his job site after working hours, the company could be sued and found liable for damages awarded to the plaintiff. If the insurance coverage is not adequate to cover the damages award, Raymond will have to come up with the cash, besides possibly feeling personally responsible for the injury to the child.

2.10. Partnership

An impatient young contractor may not want to wait several years before starting some large construction projects. If the young contractor has experience, knowledge, and ability, but no cash, then borrowed capital must be sought out and obtained if big projects are to be undertaken. It is not likely that a bank or other financial institution will lend the money to a sole proprietor with no assets, so the ambitious young contractor is advised to look to venturesome individual investors with cash for loan funds. Through personal contacts or ads in the newspapers it is possible to find private individuals who will act as sponsors or lenders and who will lend several thousand dollars at a competitive rate of interest, if there is reasonable cause for confidence in the builder's eventual success. However, many sponsors will want a "piece of the action." In other words, they will require a share in the future profits and capital growth of the company and may possibly even want to take an active part in helping run the business. Under such conditions, the more appropriate company structure would be either a partnership or a corporation.

2.11. How Many Partners?

A partnership can consist of any number of people. In most partnerships every one of the partners has some job to do to help the business prosper. In this case they are all *active partners*. In some partnerships one or more of the partners take no part in the operation of the business and are partners in name only. In this case they are *silent partners*. Usually a partner who contributes nothing but cash will be a silent partner. If the financing partners know little or nothing about construction they might prove to be a handicap to operations and the company will benefit if the financiers will accept the silent-partnership role. Yet the financier's money may provide the funds that are essential to the survival of the firm, at least while the firm is young and growing and has not yet built up liquid assets of its own to any substantial degree. If there is no agreement to the contrary, the partnership automatically becomes an equal partnership, wherein every partner has an equal share in the rewards, losses, and liabilities of the company, in a manner similar to that of the individual ownership. However, it can be stipulated that all partners are not equal, even though the partnership remains a general one. For example, one person might be a 10% partner, while each of two others has a 45% share of the business.

2.12. Liability of a Partnership

Regardless of the percentage share owned by each partner, each and every partner remains liable for all debts of the company. For example, if the partnership lost $100,000 on a construction contract and all of the partners except one disappeared, that one remaining partner is responsible for the entire $100,000, even if he only owned 10% of the firm. And, like the sole owner, the debts of the firm can reach back to the personal assets of the individual. The advantages and disadvanges of a partnership operation are therefore quite similar to those of an individual operation. The increased ease of capitalization is a big advantage for the partnership form of business, but the unlimited liability for the actions of partners over whom little or no control can be exercised is a counterbalancing factor.

One example of a very large partnership is the Van Sweringen Brothers Company, which at one time owned a railroad, a large metropolitan office building, and a limited access inter-urban streetcar line, and was the original owner and developer of the entire city of Shaker Heights, a suburb of Cleveland, Ohio.

2.13. Limited Partnership

A limited partnership is somewhat similar to a general partnership, except there are two classes of partners: (1) the limited partners, who are investors contributing funds in hope of returns, and (2) the general partners, who make policy and manage the firm.

2.14. Liability of Limited Partners

Each limited partner is financially liable only to the extent of his financial investment in the company. Suppose, for example, that a company is started with six limited partners, and each of them contributed $10,000 for a total of $60,000. In case of failure or bankruptcy of the company, or some court suit which is settled against the company, the total loss will be limited to the original $60,000 already contributed plus whatever other assets the limited partnership may have acquired since its founding. In other words, no one of the six limited partners will have to dig into a personal bank account or other personal assets, such as stocks or bonds, to cover the indebtedness of the company. The only loss will be the original cash investment of $10,000 each. Of course if the company is successful up to the time the catastrophe occurred, and has accumulated $1,000,000 in assets, it could lose all of that. However, each individual would lose no more than the original contribution already paid in to the limited partnership.

2.15.   *Authority of Limited Partners*

The limited partners are prohibited from participating in management or policy making for the company. Only the general partner can perform management and policy-making functions, such as determining what to buy and what to sell, whom to hire and whom to fire. Obviously the general partner needs to be selected with great care, since the limited partners, by necessity, put a lot of trust in the general partner.

2.16.   *The General Partner*

In a limited partnership there must be one or more persons who are the general partners. The liability of the general partners is not limited. In fact, their liability is unlimited in a manner similar to that of the sole owner of a company. In some states an individual, who is to be the general partner, may gain protection against unlimited liability by means of a corporate shield. Under these provisions, a corporation may be formed of which the general partner is the sole owner, and this corporation then assumes the role of the general partner. All the other limited partners can be individuals. There are certain special laws applying to limited partnerships, and a competent attorney should be consulted for assistance in establishing such a company structure.

2.17.   *Shares in a Limited Partnership*

The shares in a limited partnership can be divided in any reasonable way, just as the shares in a general partnership can be divided. For example, with seven partners, the six limited partners might each have a 7% share for a total of 42%, while the general partner has a 58% share. Of course, the general partner makes all management decisions, including the sale of the company's assets, regardless of how many shares the general partner holds. In the preceding example the shares could be split. Instead of 7 shares (out of the total of 100 shares) selling for $10,000, 3 ½ shares might be sold for $5000 so that instead of six limited partners and one general partner, there might be 12 limited partners and one general partner. In that case each of the 12 would have 3 ½ shares, and the one general partner would retain 58 shares. Or there could be eight partners with 3 ½ shares each, two with 7 shares, and one with 58 shares. In other words, shares in a limited partnership are somewhat similar to shares of stock in a corporation and can be subdivided and sold in any manner not contrary to the partnership agreement nor the laws of the land. Of course, the whole point of selling shares is to raise initial capital or to increase capitalization. Before selling the shares, the partners must decide on the price of each share, based on the amount of cash the limited partnership needs to raise in order to conduct its business. If they need to raise $70,000 and decide to sell only seven shares, then the "issuing price" (sale price when the shares are first issued by the partnership) will be $70,000/7 = $10,000 per share. The number of potential investors with $10,000 to invest is much smaller than the number with say $1000, so the firm might do better to issue 70 shares at $1000 per share. If the price per share is too high, all the shares will not be sold. As a result, the initial capital raised may not be sufficient to finance the proposed business. On the other hand, if the capital is raised by the sale of a large quantity of individual shares sold at a low price per share to a large number of limited partners, a great deal of extra paperwork can be expected. There will be an extraordinary amount of record keeping required as shares change hands, additional correspondence and phone calls answering inquiries and complaints from the large number of partners, and doing the general busy work that inevitably accompanies large groups of shareholders from a wide variety of backgrounds and differing outlooks. The sale of shares is subject to the rules of the Securities and Exchange Commission (SEC), a federal agency, as well as some state laws, and the counsel of a competent attorney is recommended.

2.18. *Limited Partnerships Can Provide Tax Shelter*

One advantage of limited partnerships, in addition to limited liability, involves the application of tax laws to real estate investments. The owner of real estate normally can recover the value of the buildings by means of a depreciation allowance over a 15-year period using either Straight Line Depreciation or an approved Accelerated Capital Recovery System (discussed later). During the first few years of ownership this accelerated depreciation allowance may exceed the net rental income. The result is a theoretical net "paper" loss, which constitutes a tax shelter for any outside income up to the amount of the excess depreciation. This tax shelter, plus any other profits, can be disbursed directly to each of the partners in a pro rata form; a 7% partner would get 7% of the tax-shelter benefits plus profits, and a 58% partner would receive 58%. Suppose, for example, that a limited partner has a personal income from other sources of $100,000 per year, plus a share of the cash flow from the limited partnership amounting to an additional $10,000. Now suppose that the pro rata share of the net tax-shelter loss of the limited partnership is $15,000. Then the limited partner would receive the outside income of $100,000 (personal) + $10,000 (limited partnership profit) = $110,000. This ordinarily would be taxable income, but the tax-sheltered loss from the limited partnership may be deducted from it, giving $110,000 − $15,000 = $95,000 taxable income. If the limited partner is in the 50% tax bracket, the tax is reduced from 50% of $110,000 = $55,000 down to 50% of $95,000 = $47,500. Thus the $15,000 worth of tax shelter is worth $55,000 − $47,500 = $7,500 cash to someone in the 50% tax bracket. However, if eventually the depreciated property is sold for more than the depreciated book value (it usually is), all the tax-sheltered amounts are subject at the time of sale either to recapture at regular income tax rates, or capital gains treatment at lower rates.

2.19. Forming a Corporation

Another popular method of setting up business in a form that facilitates raising capital is to form a corporation. To raise the needed capital the corporation sells shares in the ownership of the corporation. The buyers of the stock become part owners of the corporation and are called stockholders, or shareholders. A corporation may issue as many shares of stock as it cares to, subject to the approval of the board of directors and certain legal requirements. The stockholders of a corporation vote in an election, usually held at a stockholders' meeting, to elect a board of directors. Stockholders normally are entitled to one vote per share, so the holders of the majority of the stock control the corporation. The Board of Directors is responsible for formulating policies for the firm, including the hiring and firing of the president. The president as chief executive of the firm then hires the rest of the firm's employees and operates the firm in accordance with the policies of the board of directors. Of course, the services of a competent attorney are recommended when setting up a corporation and before any shares are sold. The laws regarding corporations are rigorous, and offerings for sale of shares must be handled in accordance with the rules of the Securities and Exchange Commission. Use of a corporate structure has some attractive advantages, particularly with regard to financing for a young construction or development firm.

2.20.   To Raise More Money, Just Sell More Stock

Corporations and limited partnerships need not sell all the stock at the initial offering. For example, if the original price per share is set at a value so that 24% of the total shares will produce the required initial capital, the company can retain the remaining 76% of the stock in reserve, or the founders and originators of the company can keep these shares in their own names. After a time, if the company prospers and the stock value increases, another 24% can be sold on a second offering but at the higher price. This sale should result in a substantial inflow of vital cash to increase the liquid-asset position of the company as well as that of the person who has built the company. Yet the individual who started the company can still retain control of it with the remaining 52% of the stock still held.

2.21.   Control of the Corporation

With one vote per share, the owner of the majority of the stock holds unquestioned control of the company since this owner holds the majority of votes at stockholders' meetings. If desired, this majority stockholder can move easily into a position to make all management decisions, including the sale of the assets of the company. However, the rights of minority shareholders must be kept in mind since these rights are protected to some extent by law and cannot be overlooked. There are many instances of companies being controlled by holders of less than a majority of the outstanding stock where the majority of the shares are widely distributed and in small lots and the majority shareholders have difficulty in reaching a consensus. This minority control is usually possible when shareholders of small lots are widely dispersed. Being out of touch with the corporation and not fully aware of the issues, they often do not attend stockholders' meetings, or abstain from voting their shares. In addition, the controlling minority shareholder may persuade absentee shareholders to proxy enough votes to constitute a majority of those shares being voted.

2.22.   Liability of Shareholders

One advantage of a corporation lies in the limitation of liability of shareholders. Every stockholder of a corporation is protected against any financial loss exceeding the investment already made to purchase the share of stock. The only loss that can come to any owner of shares in a corporation is limited to the loss of the money already paid to buy the shares in the company. Again, if the corporation grows to the point where it has accumulated $1,000,000 in assets after a small beginning, in the event that all those assets disappeared, no one could come after any stockholder of the company because of the corporate debts. For this reason, when some small corporations with marginal financial resources seek loans, the lenders may require one or more officers or major shareholders of the corporation to personally endorse loan notes to provide additional security for the loans. A personal endorsement involves an agreement to be personally liable for the amount of the indebtedness in case of default by the corporation. However, the loan endorsers would be liable only for those loans on which they signed, and would not be liable for other debts incurred through the daily operations of the corporation.

2.23. Priorities of Payment upon Liquidation

A stockholder is a part owner of the firm, and therefore in time of financial trouble stands last in line for distribution of the assets. If the firm is forced to liquidate, the government tax claims usually are paid first, employees salaries next, specific lien holders, commercial debts, bonds by class, and finally the owners (stockholders) divide up what's left, if anything. Each class of creditor is paid in full before the next class gets anything.

2.24. Raise Money without Selling a Share of Corporate Ownership—Sell Bonds

Bonds represent another avenue whereby a corporation can raise the capital needed for carrying on the business of the firm. A *bond* is like a note or an IOU in that it is a promise to pay interest and repay principal at a specific future date. The bond may be secured by a specific piece of property, specific revenue sources, or just faith in the company. Bonds have two basic advantages. The share of ownership is not diluted by selling bonds as it is by selling stock. Bonds represent borrowed money, whereas stock represents the sale of an ownership share of the company. The second advantage to issuing bonds rather than borrowing one large sum from a bank or other lending institution is that a corporation can borrow directly from a large number of small individual investors by issuing bonds in smaller denominations, thus cutting out the institutional middleman. The bonds serve as evidence of indebtedness. A bonded debt is scheduled for repayment at some specific future date as much as 30 years in the future. Meanwhile interest is earned at stated intervals, usually every 6 months or every year. Some bonds pay the interest payments in periodic cash payments, (usually every 6 months) and some instead credit the accumulated interest payments to the bond principal. Then at some named specific future date, called the *date of maturity*, the principal and all accumulated interest are paid in one lump sum.

2.25. Corporation Lending Rates

Another advantage to a corporate structure stems from usury laws. In many states the maximum interest rate that legally can be charged to an individual for a business loan is less than that which can be charged to a corporation. For example, disregarding small loan companies and time payments on merchandise, the maximum interest rate that can be charged to an individual might be, say, 18%, whereas for a corporation it might be 22%, depending on state laws. Since the points discount customarily charged to obtain a loan might bring the true rate of interest on a construction loan up to 19% or 20% (which is more than the 18% maximum for individuals but less than the 22% maximum for corporations) a commercial bank would not be able to make such a loan to an individual without dropping its interest rate. Because the bank feels that it needs to earn the full 19% or 20%, it will require that the contractor or developer have a corporate business structure. Where conditions such as this are in effect, the bank legitmately could make such a loan only to a corporation at the full market interest rate.

2.26. Joint Ventures

It is possible to form companies and raise capital by joining together various combinations of the four types of company structures just described into a joint venture (sometimes referred to as a "syndicate"). For example, several individual construction companies could band together into a joint-venture partnership for the purpose of taking on a particular job too large for any one of them to handle alone. The company so formed could be a partnership, limited partnership, or corporation, with each of the individual companies having a mutually agreed upon share. Remember that a limited partner contributes money only and does not participate in the operations. So limited partners in a joint venture usually would be restricted to investors or financial institutions. Each of the individual firms would retain its own company characteristics, so that the joint-venture company might be composed of two corporations and one limited partnership, and those three together would form a new corporation with temporary life for the duration of the construction project. The combined corporation could also be made up of three separate partnerships, or any other combination. The entire organization usually is disbanded and dissolved at the end of the job, when each of the member companies would again go about its own business. The assets of the joint venture can be appraised and distributed equitably to the partner firms, or sold (liquidated) at public sale or auction (with a subsequent distribution of the cash proceeds to the joint venture). The purpose of a joint venture is simply to raise sufficient operating capital and combine enough other company resources (such as experienced personnel and specialized equipment) to take on a large job without straining or exceeding the resources of any individual member. Of course, legal advice is recommended before reaching decisions on this form also.

2.27. *Separate Corporations under Common Ownership*

Sometimes a single company or person will own several separate subsidiary corporations, not joined together in a joint venture. This often occurs in a design-build operation, which provides engineering service, architectural service, and construction service to any outside owner. Under common ownership, one company designs the project while another builds the project for some client–owner. One of the reasons for doing this could stem from a desire to protect one subsidiary company from the debts and liabilities of the other subsidiary company. There could be, for example, a bid made too low by the construction corporation resulting in heavy losses and consequent liabilities. If the construction corporation should go bankrupt, the design firm would not be seriously affected because it would not be liable for the debts of the other firm. This relationship depends upon points of law to be successful and should be established with the help of a competent attorney. A second reason for establishing a separate company is for simplicity of management and control. If a contractor has difficulty determining whether a particular segment of the firm is profitable or not because all costs are not fully allocated where they should be, then one possible solution is to set up that segment of the firm as a separate, independent firm, with its own management and keeping its own set of books. For example, some contractors set up a separate firm to own and maintain their heavy equipment, renting it out to jobs constructed by the affiliated firms. A third reason lies in the tax laws. The income tax on income split between two companies is less than that on the same income lumped into one company. Here again, the IRS insists on substance over form and may disallow a division of firms done for the sole purpose of trying to reduce taxes.

2.28. *Raising New Money through Refinancing*

Sometimes a construction firm wants to undertake a large project and finds that even though it has sufficient equipment, skilled labor, and know-how, it can't take on the work because it lacks cash. For example, suppose a preliminary financial review of the proposed project indicates that the amount of cash that will be needed in addition to the construction loan is more than the company has in its bank account. The company needs to raise more money. The firm's owners may decide against taking on any more partners, and against selling any more stock from the company treasury, for either course of action would mean that each of the present owners would own a proportionately smaller share of the company. In this case, if the existing owners are reluctant to contribute more of their own funds to shore up the company finances, they may be able to raise the needed cash by refinancing some of the firm's real estate assets. On a typical existing amortized mortgage, each periodic payment is split into two parts. The first part goes to cover the interest due, while the remainder of the payment is applied to pay down the principal balance of the mortgage. As each payment is made the mortgage balance gets smaller and smaller thus significantly reducing the amount of indebtedness. Meanwhile, if new construction costs have risen along with normal inflation, the market value of the property has probably risen proportionately. Therefore, the owner's equity is represented by the growing spread between increasing market value and decreasing mortgage balance, and this owner's equity has grown. Under these circumstances, the lender of the mortgage would very likely lend the borrower an amount even greater than that which has been paid off, thus bringing the mortgage back up to a new level even higher than the original mortgage. To the borrower this means a sizeable lump sum of fresh capital. The process just described is called *refinancing*, and an existing construction or development company frequently can refinance whatever real estate property that it has held for a reasonable length of time, say 4 years or longer.

One outstanding tax benefit of refinancing is that normally the cash derived from such a transaction is tax free, because it is borrowed money. It is not money derived from earnings or from profit and therefore is not taxable except a small amount for documentary stamps. For instance, assume that 10 years ago a firm had property worth $1,000,000 and took out a first mortgage for 80% of that value or $800,000. Today the property has likely increased in value to say $2,000,000 and at the 80% loan-to-asset ratio would be eligible for a loan of $1,600,000. Meanwhile the loan has been paid down until the principal balance is only, say $600,000. Thus the firm could borrow the $1,600,000 and simultaneously pay off the current $600,000 principal balance of the earlier loan and have $1,000,000 of new borrowed money. That $1,000,000 loan would flow in tax free. Of course those tax-free dollars eventually must be repaid to the lender with interest, but interest on real estate mortgages is typically lower than on many other kinds of loans.

2.29. Summary

To start even the smallest company requires a certain amount of initial capital. In order to undertake any construction operations, some money has to be available to begin with. The amount of money needed depends on the size of the projects that the firm decides to undertake.

There are four types of company organizations, each providing different opportunities for raising capital funds.

1. An individual sole proprietor can form a company and raise money on his own, by saving, personal borrowing on his own reputation, and credit from friends, relatives, or institutions. Liability is unlimited and extends beyond the company's assets to the personal assets of the owner.

2. A group can get together, in a partnership, each contributing their own cash (or signing personally for loans) and pooling their experience, ability, and other resources. As each partner profits from the work of the others, each partner is also responsible for the mistakes or miscalculations of the others.

3. A limited partnership can be formed, in which only one or more of the partners is the general partner who accepts all financial, managerial, and policy-making responsibility, while the other limited partners (who contribute funds in hope of return and cannot participate in operations) have their loss limited to the actual cash investment.

4. A corporation can be formed, which raises part of its needed capital through the sale of equity stock (or nonequity bonds) and limits any stockholder's liability to the initial investment purchase price of the stock.

Each of the four forms of company organizations—individual ownership, partnership, limited partnership, and corporation—has advantages and disadvantages with regard to borrowing money. The methods of borrowing are similar for each. If a company's financial stability is marginal, the chief executive officer of the company, or some financially stable member or friend with adequate liquid assets, may be required to sign personally on any company note. During the course of construction large sums of money are changing hands, and the bank wants to make sure that the contractor will use the funds for the intended purpose and not divert them to pay old debts or use in some other way. If the project loses money, then the one who personally endorsed the loan note would be personally liable for seeing that the bank is fully repaid. This loan endorser has a special interest in overseeing the company operations, making sure that all money is properly accounted for, all loans promptly repaid, and the affairs of the company are properly managed.

The personal liability discussed above applies to all short-term financing such as front money, construction loans, borrowing on the contract to meet payroll, and gap financing. After construction is completed, the mortgage lender has a completed project as security and no longer needs the individual liability.

In every instance the company structure should be established only with the counsel of a competent business lawyer. The value of competent legal advice cannot be overemphasized. On the other hand, it should be recognized that attorneys are human too and can make mistakes. The attorney and the builder/developer need to work together closely and review one another's work to be sure the desired work product is achieved. The resulting details of organization always should be agreed upon in writing, particularly when two or more people are involved, as would be the case with a partnership, a limited partnership, or a corporation. Early use of a competent lawyer's expertise in the planning of a company structure can avoid future difficulty and can save many dollars as well as much time and anxiety.

3

Selecting the Market,
Choosing Sites for the Projects,
Securing the Sites

3.1. Selecting the Market

Every builder or developer produces a product or service designed to meet a particular market need. Selecting the market in which your firm can compete most successfully is an important decision when forming a new company. To improve your chances of success it is wise to select a market that is not already saturated by the competition, a market that needs a product you can produce at a competitive price with good quality and workmanship, a market that is stable and shows signs of future growth and improvement.

An aggressive builder/developer should select from the widest possible geographic market area since the wider the selection, the greater the opportunity to find superior building sites at reasonable prices. If the developer's interest extends outside the limits of the home town, then a circle should be drawn on a map encompassing the proposed market area. Probably every urban area within that circle needs some kind of construction, but whether or not it needs apartments, shopping centers, or industrial buildings depends on the current status of each local market. A good indication of current market needs can be found by determining the present vacancy rate, current construction activity, and a forecast of growth in demand. For instance, for the residential market, if the vacancy rate is less than 5% in all apartments in the area (less than 5% of all existing apartments are vacant), there may be a demonstrated need for additional residential rental units unless a large amount of apartment construction is already on the drawing boards or underway, or the population count is declining instead of growing. Furthermore, only firm rentals should be counted in the market survey. Don't include rentals where landlords give a free bonus for signing a lease, such as 2 months' free rental included in a 1-year lease, or any other gimmicks. If all rentals are firm and without give-aways then a *hard market* prevails. Given a 95% or better occupancy ratio and a hard market as described, it is still necessary to determine the type of unit to build. In an industrial town the factory blue collar workers normally can afford approximately 1 week's pay per month for housing. The project may therefore be geared to a lower rental than it would be in a prosperous white collar commuter suburb. The lower-cost units usually should be built to a higher density resulting in lower unit costs for land, paving, utilities, and other site work. Similarly the market for middle-income and luxury units has desirable special features and should be undertaken only where the economics of the local market warrant their erection. Furthermore, if all the units in the market survey area are luxury units and few lower-rent units are available, there may be a market for the low-rent units, even given a poor occupancy rate on luxury units.

3.2. Importance of Location

After determining the product (for example, apartments) and the market area (what part of what town), the next big task is to pick a site. Most real estate experts agree with the good-humored philosophy that the three most important criteria for the success of a project are: (1) location, (2) location, and (3) location. The term "location" includes most other factors of importance in selecting a suitable site for a proposed project. When considering alternative sites a builder/developer should be fully convinced of the suitability of each proposed site. Any sites that do not warrant this conviction should be discarded out of hand.

3.3. Site Selection

The goal of good site selection is to choose and secure the best available parcel of land for the type of development to be erected. The following outline of "how to select a site" uses an apartment project as an example, but with slight modifications the approach is equally applicable to many other types of developments, such as an industrial park, an office building complex, a shopping center, a warehouse, or a motel.

3.4. Factors Affecting Site Desirability

The various factors that make one site superior to alternative sites within the same area are all relative to the selected market, be it prestige, white collar, or blue collar. They include the following:

1. Location
2. Adaptability
3. Accessibility
4. Transportation and commuting time
5. Shopping
6. Schools and churches
7. Recreation

Not all of these factors will always apply to all projects, and for each project the relative importance may vary. However, to guard against forgetting or overlooking an important factor, each factor always should be at least considered. Then any nonapplicable factors may be discarded. A few additional factors may apply to some out of the ordinary sites. These could include such items as public improvements and environmental amenities. When considering public improvements, pay particular attention to paving, sewers, municipal water, and gas/electric lines servicing the property. Environmental amenities include scenic views and proximity to ocean, lake, mountains, desert, or even some man-made point of great interest, such as a cathedral. These environmental amenities constitute intangible but real influences in the general frame of location. At the same time, negative influences also should be taken into account, since one serious negative factor can offset several positive ones. Consider noise levels, odors, crime and vice activities, and the general reputation of the area. Areas with even one serious negative factor, or areas that for any reason appear to be on the decline or on the brink of decline, should be avoided as poor risks.

3.5. *Adaptability*

Although the *location* may be fantastic: for example, it is 400 yards from an expressway and well sited on a mountainside, with a clear trout stream running through it, while overlooking an 18-hole championship golf course with the ocean beyond, it may still be worthless to the developer if it is not developable or in any other way not adaptable to the project. *Adaptability* includes deed covenants, zoning and land use laws, and natural geography. Research for adaptability includes examining *deed covenants* or *restrictions*. These documents can be found in the abstract of title or in the deed, and are recorded at the County Clerk's office. If, for example, the land is restricted by deed to the building of single-family homes then no one can build apartments, even if the zoning is favorable. Covenants against race or religion are no longer valid by court decree, but covenants on minimum areas of dwelling units and maximum heights can be enforced. If these covenants are violated any property owner in the same subdivision can bring suit in civil court and have the project enjoined and stopped. Zoning laws and building codes also control adaptability and are found in local and state ordinances. These laws are passed for the benefit of the community as a whole to prevent potential damage to the value of surrounding properties caused by spot zoning or incompatible land use. Otherwise promising developments are frequently prohibited by zoning laws designed to protect the value of existing homes and other property. For example, a desirable piece of land might appear to a developer's eyes as an ideal location for a highrise building, but local ordinances might limit the height or density (number of units to the acre). Or building within a fire zone (usually a high density downtown zone with special restrictions) might require methods and materials of construction that are prohibitively expensive. Or requirements governing setbacks, side yards, and off-street parking might combine to make a proposed walk-up apartment project unfeasible. The seasoned developer does not purchase land with the optimistic (and usually unfounded) hope that the zoning restriction against the expected use can be changed. The process of zoning change is long and arduous and usually unsuccessful. Even if the zoning can be changed, the time and expense involved often make the effort infeasible if alternative sites are available.

Finally, the term *geography* embraces both topography and subsoil, both of which are significant influences on construction costs. If a visual inspection of the site shows swampy or marshy land, steep ravines, rock outcroppings, or poor surface drainage, the site preparation probably will require a significant amount of work and expense before any buildings can be started, thus in effect increasing the land acquisition cost for the project. Similarly, if preliminary inspection reveals a high water table, or pockets of muck underneath the topsoil, the construction cost probably will be increased. To compensate for the increased construction cost the seller of the property should be expected to accept a correspondingly lower selling price for the land purchase. Excessive excavation and other site preparation costs should not come as a surprise after financing is arranged. If any doubt exists regarding the cost of dealing with the geography of a site, the land should be placed under option-to-purchase until test borings or test pits reveal the true nature of the sub-surface condition. Test borings are not expensive and competent testing laboratories are widely available.

3.6.   Accessibility

An otherwise fine piece of land is worthless if there is no easy way of getting to it, or if access is unreasonably restricted or otherwise inadequate. In most cases potential renters or buyers look for a place with easy access and like to drive automobiles right up to the front door. Good locations for residential use should have frontage on a local street, a short distance from an arterial. Of course it is possible to build successful apartment projects alongside six-lane expressways providing that proper buffering is installed. *Accessibility* does not preclude building out in the country, provided that a major highway passes nearby. Most people (except in Manhattan and a few other exceptional places) habitually rely on the automobile as their major mode of transportation, and inadequate access should be shunned. For example, access through a blighted run-down area or by way of a stretch of dirt or congested road greatly diminishes the desirability of a site. Of course there will be exceptions. For instance if the project is within a short, pleasant walk to work, shopping or mass transit, it could prove to meet the needs of a significant rental market even without good accessibility for autos. The meaning of "accessibility" will vary somewhat depending upon the particular market selected and a common sense evaluation is advised.

3.7. Transportation

While nearby public transportation (bus, train, subway) on frequent schedules can be one important means of providing accessibility, both public and private transit are considered in computing *commuting time* from the apartment to work. Either private or public means of travel may be used, or the two may be joined, but in any case, the resulting *time* from dwelling to work should not be excessive. The travel time usually is considered of greater importance than travel distance because people generally regard time as the governing factor and don't mind traveling longer distances if it can be done at higher speeds within a reasonable time span. There is usually a greater selection of desirable dwelling sites further away from the employment center. A 10-mile drive on an expressway may actually take less time than a 2-mile ride on a local bus, and there is a much greater selection of desirable sites within a 10-mile radius than within 2 miles. In some areas bicycling is coming into favor, so accessibility to bikeways also should be considered.

Public transportation assumes greater importance in densely populated urban areas. Even in low density urban areas public transportation is helpful to many potential renters and is necessary for some. Since bus routes are flexible and can be routed to where potential riders live, a large enough rental project could generate an extension of the bus route out to the new development. Of course, the prospect of such extensions cannot be relied on, but could serve to further enhance the value of an already feasible project. An ideal site will be located within a short walk of existing bus or commuter train lines on routes as direct as possible between home and work so that executives do not have to make transfers on their way to and from the office.

3.8. Community Facilities

The availability of *community facilities* should also be considered. These include shopping, schools, churches, and recreation such as parks, playgrounds, and entertainment centers. It is preferable to be within walking distance of such community facilities, but this is not mandatory. With a good transportation network available, and easy access by automobiles, physical proximity is not essential although it is desirable. Short travel times and low travel costs are the real criteria for determining "closeness" to community facilties of all types.

3.9. Rating Each Site by Matrix Analysis

Each of the factors so far presented can be given a weighted value according to its order of importance. Each weighted value is a percentage, and the total of all the percentages should be 100. One suggested list of weighted factors follows:

Factor	Percent
Location	25
Adaptability	20
Transportation and commuting time	20
Accessibility	15
Shopping	10
Schools, recreation	10
	100

Alternative sites can be compared by rating the factors at each site. First consider each site separately and grade each factor at the site with a grade from 1 for extremely poor, to 100 for perfect. Then multiply that grade by the appropriate percentage opposite each factor and find the number of points for that site, for that factor. For example, if at Site 1 the location factor was given a grade of 88, that site would earn $88 \times 0.25 = 22$ points for location. If Site 1 had a grade of 85 for transportation and commuting time, it would earn $85 \times 0.20 = 17$ points additional. Adding up all the points for each site would lead to a relatively unbiased selection of the best location, instead of a selection based completely on instinct and intangibles such as someone "just loves that darling oak tree." An example follows:

Factor	Value (V)	Site 1 Grade/ $G \times V$	Site 2 Grade/ $G \times V$	Site 3 Grade/ $G \times V$
Location	0.25	88/22	72/18	60/15
Adaptability	0.20	80/16	70/14	80/16
Transportation and commuting time	0.20	85/17	65/13	75/15
Accessibility	0.15	80/12	80/12	90/14
Shopping	0.10	70/7	80/8	70/7
Schools, Recreation	0.10	80/8	70/7	60/6
Total	1.00	/82	/72	/73

Obviously Site 1 is the clear-cut winner with 82 points. But if Site 1 becomes unavailable for some reason, the choice between Sites 2 and 3 is made difficult by the closeness of the score between the two (72 points for Site 2, and 73 points for Site 3). Even though Site 3 scores 1 point more than Site 2, the location superiority of Site 2 (18 points vs. 15 points) might be reason for reconsideration. A difference of only 1 point in 100 points is not significant and for practical purposes the sites should be rated as about equal.

The grade points based on values that are given to each of the six different factors constitute an approach called *matrix analysis*. The various points should not be based on hunches or gut reactions or even emotional preference, but should be objectively determined by studied consideration of data. The data can be put in any form that seems suitable, but they must be based on actual field inspection reports. A suggested check list for site appraisal is given in Figure 3.1 as one method of obtaining relevant data. In addition to filling out such a seven-page form, it is advisable to prepare a location map showing the topography, and to photograph the site and its neighborhood amenities to verify visual impressions. Additional check lists for commercial projects are given in Appendix A.

3.10. Reserving the Land by Option

Once a promising site is finally located, the site should be reserved by option so that no one else can buy it while the financing and other preliminaries are being arranged. Standard option agreement forms are available which provide that the owner of the land gets a sum of money that binds the owner to selling or leasing the land to the buyer, if the buyer so desires, and prevents the owner from selling (or otherwise encumbering the land)

CHECK LIST FOR SITE APPRAISAL

Site _____

Date Visited _____   By _____

Comments: _____

_____

OWNERSHIP AND DESCRIPTION

Owner _____

Agent _____ (address) _____ (phone)

Address of Tract _____ (address) _____ (phone)

Legal Description _____ (street) _____ (city) ____ (county)

_____

Tract Size: Dimensions _____ X _____ Acres _____

Adjacent Land Owner _____

Comments: _____

_____

LAND USE AND VALUES

Asking Price: Per Acre _____ Total _____

Debts Against Tract _____

    Held by _____

Special Assessments _____

Buildings on Tract _____

    Est. Value _____

Land Use Now _____

Tax Assessor's Value _____ Tax Rate Used _____

Your Estimate of Value: For Use Pending Development _____

    For Development _____ Per Acre _____ Per Lot _____ Total _____

Comments: _____

_____

FIGURE 3.1. Check list for site appraisal.

TOTAL LOCATION

Miles From City _____ Direction _____

Nearby Communities: _____

Distance From Central Business District: Miles _____ Minutes by Car _____

    Minutes by Public Transportation _____ Fare _____

Comments on Highways _____

Type of Public Transportation _____ Good ☐ Fair ☐ Poor ☐

Local Streets and Roads _____ Good ☐ Fair ☐ Poor ☐

Approach to Tract _____ Good ☐ Fair ☐ Poor ☐

Surrounding Development: Class; High _____ Medium _____ Low _____

    Price Range of Homes _____

Employment Centers: _____

    Type _____ Miles _____ Minutes _____ Fare _____

    Type _____ Miles _____ Minutes _____ Fare _____

    Type _____ Miles _____ Minutes _____ Fare _____

Remarks: _____

_____

_____

ENVIRONMENT AND CONTROLS

Zoning of Tract _____ Nearby Lands _____

_____

Protective Convenants for Tract _____

    For Nearby Lands _____

Subdivision Regulations _____

Any Easements on Land? _____

Subdivision Plan Recorded? Yes ☐ No ☐; Area Master Plan? Yes ☐ No ☐

Comments: _____

_____

FIGURE 3.1. *(Continued)*

HAZARDS AND NUISANCES

Is the Tract Free From Existing or Likely:

	Yes	No	Remarks
Heavy and Frequent Rail Traffic?	☐	☐	_____
Heavy Highway or Street Traffic?	☐	☐	_____
Airport Noise and Hazards?	☐	☐	_____
Aircraft Approach Patterns?	☐	☐	_____
Other Unusual Noise, Vibration?	☐	☐	_____
Unusual Crowds, Heavy Parking?	☐	☐	_____

Is There Adequate Distance and a Buffer Between the Tract and: _____

	Yes	No	Remarks
Likely Sites of Fire?	☐	☐	_____
Smoke Sources?	☐	☐	_____
Chemical Odors?	☐	☐	_____
Other Odors?	☐	☐	_____
Dust or Dirt Sources?	☐	☐	_____
Unsightly Views?	☐	☐	_____
Floods?	☐	☐	_____
Polluted Bodies of Water?	☐	☐	_____
Dilapidated Structures?	☐	☐	_____

Observations: _____

_____

_____

_____

_____

_____

_____

_____

_____

_____

FIGURE 3.1. *(Continued)*

TOPOGRAPHY

Is the Tract Characterized by or Are There Large Acres of:

Yes No   Remarks

Swamp or Marsh? ☐ ☐ _____

Steep Ravines or Grades? ☐ ☐ _____

Rock Outcroppings: ☐ ☐ _____

Soil Erosion? ☐ ☐ _____

Rocky or Sandy Soil? ☐ ☐ _____

High Water Table? ☐ ☐ _____

Poor Surface Drainage? ☐ ☐ _____

About how much of the tract is: Hilly? ____ % Rolling?_____ % Level ____%

What is the average elevation of the land? _____High Point___Low Point____

Are existing roads or streets above grade or below? _____

Heavy Cutting Needed: For Streets _____For Homesites_____

     For Other Purposes _____

Heavy Filling Needed: For Streets _____ For Homesites_____

For Swampy Areas _____For Other _____

In General, is the tract adequately drained? _____

Special Drainage Provisions Needed: _____

Type of Soil _____

Trees: Woods _____ Sparse Woods _____ Scrub _____ Open ____

Comments: _____

_____

_____

_____

_____

_____

_____

_____

FIGURE 3.1. *(Continued)*

UTILITIES AND SERVICES

Are These Present or Readily Available? _____

                          Yes No

Public Water Supply? ☐ ☐ Distance to Main _____ Size of Main _____

Private Water Needed ☐ ☐ Individual System _____ Private Wells _____

    Remarks _____

_____

Public Sewage Disposal? ☐ ☐ Distance to Main _____ Size, Invert _____

Private System Needed? ☐ ☐ Subdivision Plant _____ Septic Tanks _____

    Remarks _____

_____

Storm Sewers? ☐ ☐ _____

Paved Access Streets? ☐ ☐ _____

Public Street Upkeep? ☐ ☐ _____

Snow Removal & Sanding? ☐ ☐ _____

Police Protection? ☐ ☐ _____

Fire Protection? ☐ ☐ _____

Garbage Removal? ☐ ☐ _____

Electrical Utility? ☐ ☐ _____

Telephone? ☐ ☐ _____

Gas? ☐ ☐ _____

Street Lighting? ☐ ☐ _____

Observations: _____

_____

_____

_____

_____

_____

_____

_____

FIGURE 3.1. *(Continued)*

TYPE OF DEVELOPMENT SUGGESTED

Price Range of Homes: Average _____ High _____ Low _____

Number and Nature of Unsold New Houses in Area: _____

_____

Sizes of Lots Required: _____

Estimated Lot Yield: _____ Per Acre _____ Total _____

COMMUNITY FACILITIES NEEDED:
  Shops, Schools, Churches, Other Structures: _____

_____

  Parks, Swimming Pool, Playgrounds, Etc.: _____

_____

Summary of Observations on Worth of Tract Against Market Survey Results: ___

_____

_____

_____

_____

_____

_____

_____

_____

_____

_____

_____

_____

_____

_____

_____

_____

_____

FIGURE 3.1. *(Continued)*

COMMUNITY FACILITIES

Name, Address	Distance— Center of Tract	Transportation Time Minutes Walk	Car	Public	REMARKS: Crowded, Age, Condition, Other Factors
SCHOOLS					
Nursery					
Kindergarten					
Elementary					
High School					
Other					
SHOPPING					
Main Centers					
Neighborhood Stores					
CHURCHES					
Denominations					
PARKS					
PLAYGROUNDS					
SWIMMING POOL					
LAKES, STREAMS					
OTHER OUTDOOR SPORTS AREAS					
ORGANIZED SPORTS AREAS					
Golf					
Riding					
Other					
ENTERTAINMENT CENTERS					

FIGURE 3.1. *(Continued)*

to anyone else. In other words, the buyer pays for and gets, for a specified length of time, an *option* to buy or lease the land. The price of the option depends upon the terms. For instance, if farmland suitable for a proposed development is selling for $2000 per acre, but the proposed development project would actually support a land cost of up to $30,000 per acre, a farmer may well accept $1 per acre for a 1-year option to sell his farm land for say $10,000 per acre in the event a viable project can be put together. In this case the real value received by the farmer for the option is the chance to sell the land at a higher than normal land-sale value if the project goes through. The farmer takes a low price for the option in exchange for a chance at a high price for the land sale. On the other hand, if the option were written to purchase the land at a lower price of say $2000 per acre (equal to the current market price as farmland), the farmer would be justified in asking a much higher price for the option since the farmer is relinquishing for one year any opportunities to sell at current rates, and does not realize any special benefit from the project. In this case, from the farmer's viewpoint, the break-even price might be the lost interest had the land been sold and the funds invested. If lost interest is calculated at 12% on $2000 per acre, then the farmer might ask for at least $240 per acre for the option for 1 year.

At the expiration of the option period, if the buyer decides not to exercise the option to buy, the land owner keeps the option money and the land. If after obtaining the option, it is found that the project won't work to full satisfaction, the buyer must be prepared to drop the option and lose the cost of purchasing the option. For instance, if sufficient financing is not available at a reasonable rate, or if construction costs should suddenly increase, or the market collapses, or for any other valid reason, the buyer must be mentally and financially prepared to forfeit all the option money. That option money is considered an unavoidable ordinary expense of carrying on the business of land development.

4

Will It Work ?
Calculating the Feasibility/
Profitability

4.1. The Feasibility Study

After determining that at least one suitable site is available, a feasibility study should be carried out to determine what kind of a profit, if any, that the proposed project will return. A thorough and successful feasibility study is important to the owner/developer for two reasons: (1) the study assures that the owner/developer is embarking on a venture with a high probability of success, and (2) the study is needed later as part of a presentation to lenders when applying for financing for the project. The lenders will study the feasibility report to determine the soundness of the project and to be sure that their loan funds will be well secured by a viable and profitable project.

4.2. An Example of a Feasibility Study

The example shown in Table 4.1 involves a hypothetical apartment project, but the form of the presentation would be quite similar for any other real estate venture, such as a motel, a warehouse, or an office building, provided that certain modifications were made here and there as applicable. An explanation follows the example.

TABLE 4.1. Income Approach to Value for Picanos Villa Apartments, Urbane City, Anystate

Income

Number of Units	Type	Area (sq. ft.)	Rental per Month	Rental per Year for Each Unit	Rental per Year for All Units
25	Studio	500 × 83¢ = $415		$4,980 × 25 =	$124,500
35	1 BR	680 × 80 = 544		$6,528 × 35 =	228,480
18	2 BR	800 × 80 = 640		7,680 × 18 =	138,240
22	2 BR	850 × 78 = 663		7,956 × 22 =	175,032
10	3 BR	950 × 76 = 722		8,664 × 10 =	86,640
110	Total =	78,900 sq. ft. rental area		Gross rental income (100%)	$752,892

Less vacancy and collection loss at 7%	−52,702
Gross rental income (93%)	$700,190
Auxiliary income	10,000
Effective gross income	$710,190

Operating Expenses

Electricity (common areas only at $2.80 per unit per month) 12 × 2.80 × 110	$3,696
Waste and sewage at $10.00 per unit per month 12 × 110	13,200
Garbage, including dumpster rental at $7.20 per unit per month 12 × 110	9,504
Janitor and yardman	
1 full time at $300 per week 52 × 300	15,600
1 part time at $150 per week 52 × 150	7,800
Painting at $240 per unit per year 110 × 240	26,400
General repairs at $120 per unit per year 110 × 120	13,200
Supplies at $40 per unit per year 110 × 40	4,400
Replacement reserve	
Mechanical equipment	18,000
Carpeting	22,000
Management at 5% of effective gross income	35,510
Taxes	
Payroll	4,800
Real Estate	64,000

TABLE 4.1. *(Continued)*

Operating Expenses

Insurance	5,200
Advertising and accounting	5,500
Total operating expense	$248,810
Effective Gross Income	$710,190
Less total operating expenses	− 248,810
Net income	$461,380
Less interest on $700,000 land at 12%	− 84,000
Annual net income available for interest on and recapture of value of improvements	$377,380
Building value capitalized at 12% (i.e., $377,380/0.12)	3,144,833
Add land value	700,000
Value (by Income Approach)	$3,844,833
First mortgage loan at 75% of value (i.e., 0.75 × $3,844,833)	2,883,625
Loan requested	$2,883,000

4.3. Picanos Villa Apartments, Explanation of Table 4.1

The number and mix of the units should be maximized to obtain the most return from the project. The "mix" means the type of units that will be built, that is, the relative percentage of three-bedroom units, the percentage of two-bedroom units, and so forth. In this example, perhaps the builder asked his architect to produce a design containing 50% two-bedroom units and 33% one-bedroom units, the rest being no-bedroom or "studio" apartments. The builder decided on this balance after studying the local market, looking at trends in the area, and considering national trends. This ratio must be decided only after careful consideration. It constitutes one of the more important decisions affecting operating management that the developer makes in this project.

4.4. Putting the Architect to Work

As soon as the site has been selected and an option to buy has been placed on it, an architect should be engaged. Of course an architect cannot work without input from the developer. The architect will need a topographic survey showing the size of the plot, the relative elevations of the land, the location of the street and all pertinent utilities, and so on. But the developer makes the basic decisions on what type of project to build and what amenities it will contain, such as a swimming pool, laundry building, utility building, clubhouse, tennis courts, and so on. The architect also needs to be told the approximate size of each apartment unit in terms of bedrooms and square feet. This is part of the original decision-making process of the developer. The architect then puts all these parts of the program together and attempts to satisfy the client while designing a handsome building. In this example the architect came quite close to the owner's request for 50% two-bedroom, 33% one-bedroom, and the rest studio units, while placing as many units as possible on the land. Maximizing the number of units decreases the proportional cost of land per unit, and actually increases the relative amount of mortgage per unit. For certain selected markets, however, it may be desirable to disregard density as a goal and purposely cut down on the number of units that could be built. Lower density can lead to larger open spaces and perhaps to a more attractive and more rentable complex. The final decision on density depends on the cost of the land, the type of competition in the area, existing zoning regulations, off-street parking requirements, building height restrictions, and side yard and setback rules as well as what segment of the rental market is being courted.

4.5.   Rental Rates

The schedule of rental rates is the next major decision that must be made by a developer. A survey of the local market is needed before any project is begun to determine what amenities should be included, what rent per square foot is being charged against those amenities and the raw rentable space, and how the local rental trends compare with regional and national trends. If, for example, all existing apartments in Urbane City have two bathrooms, then either any proposed apartments will have to provide two bathrooms or else the proposed apartments must have a reduced rent to remain competitive if they are to have only one bathroom. On the other hand, any additional features that are not provided by the competition will call for a higher rent than that charged by others. If the competition offers simply living space but the proposed project will have a swimming pool and a clubhouse, the projected rental can be higher than that of the competition. Other amenities could well include fireplaces, dishwashers, and sauna baths, but it is well to remember that additional mechanical gadgetry demands increased maintenance, and that the increased rental might not prove to be worthwhile with these higher maintenance costs.

There is also future replacement of equipment as it wears out, which can prove expensive and troublesome. Amenities do not always pay for themselves. Their inclusion depends entirely on the market for which the project is being built.

TABLE 4.2.   Competition Comparison Analysis

Amenities	Project Identification Number							
	1	2	3	4	5	6	7	8

1. Rent ($)
   1 BR
   2 BR
   3 BR
2. Space
   1 BR
   2 BR
   3 BR
3. Monthly rent per
   square foot
   1 BR
   2 BR
   3 BR
4. Balconies and Patios
5. Kitchen Equipment
   a. Disposal
   b. Dishwasher
   c. Refrigerator Size
6. Eat-in space in kitchen
7. Number of Bathrooms
   2 BR
   3 BR
8. Air conditioning
9. Other equipment
   a. Washer and dryer
   b.
   c.
10. Flooring
    a. Parquet
    b. Carpet
    c. Other
11. Draperies
12. Swimming pool
13. Others
    a. Playground
    b. Recreation lodge
    c. Nursery school
    d. Extra Landscaping

4.6.   High-Rise Rentals

In a multistory building the costs per square foot of construction increase as the building gets higher. Therefore the rental must be more in a high-rise building than for comparable space in a one- or two-story walk-up structure. The difference will probably be at least 10% more on the average. In the final analysis, rent can be determined only by local market conditions, regardless of what the national averages show.

4.7.   Variations in Rent

Notice that in the example of Table 4.1 the rent per square foot per month is not uniform for all the different apartment units. The reason for this spread is due to the cost of construction. It costs more to build a studio apartment than a two-bedroom apartment because a studio apartment still has to have a bathroom and a kitchen, which are very expensive, and the cost of these two rooms cannot be spread out over a large area of living space. The square-foot cost of a bathroom may be as much as ten times the square-foot cost of a bedroom. Thus, when a 500-square-foot apartment unit has a 35-square-foot bathroom, and when a 1000-square-foot apartment likewise contains one 35-square-foot bathroom, the total cost of the 500-square foot unit will be considerably more per square foot than the comparable square-foot cost of the larger unit.

Similarly, the top floor of a multistory building costs more to construct than a lower floor, and therefore must yield a higher rental if each rental unit is to earn a rent comparable to its cost. It is also more desirable to be higher up—the view is better, it is quieter, and there is less disturbance from people going by. For these reasons also, a unit on a higher floor commands a higher rental.

Location within a project may also affect the rent schedule. In an office building a corner office is more desirable than an interior space, and in an apartment a unit with a better view or a unit in a prestige location will carry a higher dollar value. Thus balcony units bring more rent than ground-floor units, and those overlooking a valley are worth more than those facing a parking lot.

4.8.   Future Estimates versus Inflation

When estimating future costs and income we often ignore the effects of inflation, assuming that higher costs offset the higher income. If operating expenses rise, then rents will rise sufficiently to cover the costs. Lending institutions usually have accepted this simplistic rationale and not required any more complicated approach. They generally recognized that inflation favors the property owner by increasing the owner's equity and making the loan more secure. Therefore if a proposed project is feasible with constant, noninflated dollars, then it should be even less risky should inflation occur. Thus the examples shown in this chapter appear in terms of constant dollars.

Exceptions to using constant noninflated dollars will occur. Whenever investors believe that inflation is predictable, they begin investing with that assumption. One of the important by-products of inflation in real estate is the appreciation of property values rather than depreciation as the years pass. If inflation causes real estate prices to rise by 6% per year, then a typical project purchased a year ago for $1,000,000 is now worth $1,060,000. If the project was purchased for $100,000 down payment, then the buyer made a profit of $60,000 on a $100,000 investment (assuming that the one year's rents and tax shelters offset the one year's operating, interest, and property-tax costs). Thus in cases like these, constant-dollar costs and incomes need not break even year to year in order to make a profit upon resale at the appreciated resale price that occurs due to inflation. One unfortunate result of the use of constant noninflated dollars in the feasibility study is that some projects that actually are quite profitable and feasible, if a predictable amount of inflation were assumed, appear to be marginal or unfeasible when the study is done with constant noninflated dollars.

4.9. *Vacancy Loss*

The next item shown in Table 4.1 after "Gross annual rental income" is "Vacancy and collection loss." This is included to account for the fact that the maximum possible gross annual rental income is based on 100% occupancy for an entire year. It would be very nice if 100% occupancy could be achieved, but in all likelihood there will be some vacancies, and at least some income will be lost in turnaround time between the time one tenant moves out and the next one moves in. Furthermore, a small percentage of tenants will leave without paying a current bill, and there may be some bad checks that cannot be collected when certain tenants leave town.

When the two factors, that is, the vacancy ratio and the bad debts, are added together 7% seems like a reasonable figure to use for vacancy and collection loss. Many mortgage lenders will accept a 5% figure. In the case that we are now studying, the 5% vacancy will lead to an increase in projected income of $15,057, that is, 5% of $752,892 = $37,645 instead of the $52,702 shown, so that the gross income with a 5% vacancy loss would become $725,247. Eventually, if the new figure is followed all the way through the table of income approach to value, the first mortgage will increase by $94,106. Thus, if a 5% vacancy ratio is used instead of 7%, a higher first mortgage loan can be obtained. The 7% figure was used here to be on the conservative (pessimistic) side, but a 5% figure is commonly encountered.

4.10.   Auxiliary Income

The main business of a real estate project is the rental of space. However, there are opportunities for obtaining income from other sources, for example, from laundry machines, vending machines, enclosed parking spaces, or fees for swimming club memberships in the case of a motel in a city. It is not necessary for the project to actually own any vending machines. They could be owned by some other company that pays a royalty or fee to the project. This royalty or fee would become auxiliary income. It is well to note that the amount of mortgage money borrowed is based on the project value, which in turn is based on net income. Therefore every additional dollar of income will mean an additional $6.25 in the amount of the mortgage money that can be borrowed [($1 × 0.75)/0.12 = $6.25], and this, in turn means that the owner will not have to put in as much owner's capital—the owner's equity will be reduced by that $6.25 of borrowed money. It is therefore a good idea to get as many dollars as possible into the income column, but at the same time every dollar shown there must be realistic.

In fact the entire rental schedule must be realistic. Once the mortgage broker and the lender accept the rental schedule submitted by the developer, and the lender makes a commitment based on that schedule, the developer will be held to it. That stipulated amount of income will have to be produced before the mortgage is funded in cash. If the rents are too high, the apartments will remain empty, and the minimum floor of 81% of dollar occupancy required by the lenders to qualify for the mortgage loan ceiling amount will never be realized. The developer will perhaps then lose the project as well as the equity in it. In addition, much more can be lost. The developer who signs personally for any loans will be held responsible for their payment. Therefore the developer's decision to adopt a rental schedule must be very carefully considered, as it is a most important one. Lower rents mean a lower mortgage loan amount, and a lower mortgage loan means that the owner must provide more equity cash. That is to be avoided if possible. On the other hand, higher rents, which lead to a higher mortgage loan, might lead to so many vacancies that the owner could not meet the financial obligations, and the entire project might be lost. In summary, the leverage of $6.25 of mortgage loan money for every dollar of income produces a large swing in both directions and must be very studiously balanced. The rent must be at just the right level to maximize the profit with the greatest possible occupancy, and this rental rate will produce the maximum possible amount of mortgage money with the least owner's money.

4.11.   Operating Expenses

After a realistic total income is computed, the net operating expenses for the year must be determined. The cost of operation of a real estate project can be divided into seven broad categories:

1. Utilities
2. Payroll
3. Maintenance and repairs
4. Replacement reserve
5. Management
6. Taxes
7. Miscellaneous

1.   The *utilities* to be listed are those that must be paid by the project. These include public lights, public water and sewer, and any other utilities included in the rental. Some places pay all electric and gas bills of tenants, and others include heat and air conditioning in the rental, all of which are utility costs.

2.   The *payroll* of the project covers the yardmen, maids, janitors, and maintenance help who are employed on a regular basis and are not a part of the management of the project. The payroll often includes payroll taxes, insurance, and any fringe benefits offered by the company.

3.   *Maintenance and repairs* include all items that are the responsibiltiy of the owner. In some cases, such as shopping centers, the owner takes care of the exterior and the tenant is responsible for everything inside the building shell. In apartments or motels the owner must take care of everything that would be included in normal wear and tear, such as repairing a faulty toilet.

4.   A successful business always has enough cash on hand to cover emergencies. Real estate operations are no exception to the rule, but too many owners neglect to put money aside, preferring to operate on a hand-to-mouth basis while hoping for the best. They drain as many dollars as possible from the project in the mistaken belief that difficulties will never catch up with them, but sooner or later the day of reckoning comes. At that time the money must be found to replace thousands of yards of carpeting, or to buy 100 new refrigerators, or whatever the case might be. Therefore a *replacement reserve* should be set aside every year into a separate bank account, so that the necessary money will be on hand when it is needed to pay for a large expense of capital equipment. Such a fund should also include something for repairing or replacing the roof, for exterior painting, and so on. In any event, the mortgage lender knows that the expenditure will occur at some future date, and can easily equate that future expense to a uniform yearly cost, as will be shown in Chapters 12 and 13. The yearly cost (or savings set aside toward the future cost) must therefore be shown on an income approach to value as an operating expense.

5.   *Management* expenses are those strictly associated with the costs of management. These include the costs of collecting rents and making out checks to pay the bills, as well as the salaries of a resident agent, if there is one, and of his assistant, if he has one, plus any home office expense associated with the project.

6.   *Taxes* include licenses, fees, and ad valorem (according to value) real estate taxes, in addition to any income taxes paid by the company before any money is transferred into the hands of the owners. Some prefer to include payroll taxes in this category, but it seems more logical to put payroll taxes into item 2, since they are a direct cost.

7.   *Miscellaneous* covers everything else. It is a catch-all for anything that does not fall into the first six categories of expense items. It might include, for example, advertising, accounting, and legal costs.

In Table 4.1 these seven headings have been subdivided into more specific divisions so as to gain a greater degree of accuracy in predicting expenses. Expenses should actually be estimated as accurately as possible, and they should be based on existing costs and conditions in so far as they can be determined at the time the project is proposed, for much the same reasons as given for income prediction. If the expenses are set too high, the mortgage will not be as large as it should be, and the owner will have to invest an undue amount of cash. Every dollar that can be saved on expenses will add a dollar to the net income, because expenses are subtracted from gross income to obtain net income, and every dollar added to net income means about $6.25 more on the mortgage loan. On the other hand, if the expenses have been underestimated, the mortgage will be so large that the payments on the inflated amount may prove to be an impossible burden. Once the development is a reality, it will become necessary to meet all bills. If all the money has gone to make mortgage payments, so that none is available for maintenance, for example, then either the tenants will move out of the resulting slum or the creditors may force the owner into bankruptcy. There must be enough income to pay both the expenses and the mortgage, and the income cannot be increased without destroying a very carefully balanced rental schedule.

The total of all operating expenses, then, does not include payments on the mortgage. And this total of all operating expenses should not exceed about 40% of effective income. In the case of Table 4.1, total operating expense is ($248,810/710,190) equal to 35% of effective gross income, which is well within the 40% allowable. Therefore the figures presented should be acceptable to a mortgage broker.

4.12.   Mortgage Payments Not Included in Operating Expenses

The operating expenses do not include payments on the mortgage. Every payment on the mortgage includes both principal and interest. The part that goes to pay off the principal is not an expense but is a reduction in debt and therefore an increase in equity.

4.13. Verify All Costs Locally

All the dollar amounts shown in Table 4.1 represent the costs of a hypothetical project and no particular relationship is claimed for them. For an
actual study all costs must be verified by local professionals familiar with
costs in the area in which the project will be constructed. Costs are subject
to change not only from location to location, but with time as well. Estimates
of last year's construction costs are typically low compared to the next year's
actual construction costs.

4.14. Land Cost and Value

Like most other values, land has a value equal only to what a willing seller
and buyer agree it is worth at the time of the transaction. It can be appraised
and reappraised. It can be thought of as being extremely valuable or not
worth very much. But if a land owner sells a certain tract to a developer
for $700,000, then that parcel is worth $700,000 at the time of the sale. It
takes two people to establish value and make a sale, a buyer and a seller.
Neither of them can establish value alone. The seller can offer, and the
buyer can counter offer, but until they reach agreement on a sale the value
attributed by either is not verifiable.

4.15. Lost Interest on Land is a Cost Charged to the Project

When a developer buys land and pays cash for it, that cash can no longer
be invested and earn interest. If the interest rate on the permanent loan
is 12%, then any money connected with the project should earn 12%.
Therefore it must be assumed that cash tied up in the project could earn
at least 12% if invested elsewhere, because the going rate of return has
been established by the mortgage rate. In other words, any cash put into
the land is losing 12% interest that could have been obtained if the same
money had been invested elsewhere. Therefore, the interest on $700,000
at 12%, which comes to $84,000 per year, is deducted from the net income
of $461,380, leaving $377,380 available for interest on, and recapture of
(principal and interest payments), the value of the improvements.

The foregoing procedure is used by a number of astute mortgage brokers
when computing the income approach to value and is presented in this
example because lost interest on the land really is a cost that should be
charged to the project. The concept of subtracting the cost of lost interest
from the income is by no means universally accepted, but really should be
since that lost interest is a cost of doing business. The developer has, in a
sense, buried the purchase money in the land, and has sacrificed any div-

idends or interest that otherwise would result. The only payoff for the land investment is the income from resale. Therefore on resale the developer must try to recover not only the original purchase price but also all the accumulated interest that would have compounded if the money had been invested in some other way.

4.16. Cap Rate

The question now is, "What is the value of a business that produces a net yearly positive cash flow of $377,380?" It is assumed that this amount will be forthcoming until the project is sold for its original cost or more. In every year from now on until sold, the business should have a net cash flow of about $377,380 after expenses (except mortgage payments) are deducted from gross income. This preliminary example does not consider inflation or deflation, since we are assuming only constant-value dollars.

Now the value of the project can be found by dividing the yearly return by a capitalization rate, or *cap rate*. The value thus found is the worth of the project as an investment. It is not the value of the buildings or of the land or of any combination thereof, but is the value of the project as a business in terms of today's constant-value dollars.

There is no formula or scientific basis for establishing a cap rate, since it fluctuates in the same way that interest rates fluctuate. During a period of inflation, or at times when there is a high demand for money, the cap rate will rise along with other interest rates. This will decrease the apparent value of the project, which is equal to the uniform annual net income divided by the cap rate. The larger the denominator, the smaller will be the resulting value:

$$\text{Investment market value} = \frac{\text{Payout}}{\text{Cap rate}}$$

A smaller denominator will produce a larger investment value. The cap rate is set by the lender and may differ from one lender to the next. A high cap rate indicates that a lender's funds are scarce and they are not very eager to lend money at that time. A lower cap rate value results in a lower mortgage and a higher equity. The lender is thus asking for more security, and less risk for the loan. Conversely, when the demand for funds is weak, and the supply of money is plentiful, the cap rate will usually drop. The lower cap rate results in a larger value for the proposed project, so that a lender can lend out more money and invite a developer to borrow funds immediately instead of waiting.

The cap rate in effect at any point in time is completely at the lender's discretion. There is no argument about it. Only a mortgage broker or banker can say what different banks are using for their respective cap rates at any specific time. If the cap rate is high, the borrower may postpone the project, or find a lender willing to set a lower figure.

4.17. *Value of Income Approach*

In any event, with a cap rate of 0.12, the building value becomes $377,380/ 0.12 = $3,144,833. This is not the market value or construction cost of the buildings, but rather the value of the investment. It is considered to be the value of the buildings only for the purpose of computing the amount of the mortgage loan. The buildings themselves, it should be remembered, are not worth anything to the lender. The lender is lending money to a business. The actual cost of the buildings could well be less than $3,144,833. But even if the buildings were completely leveled by fire or earthquake, the land will still have a value of $700,000. Therefore the total value by the income approach is the sum of the business value of $3,144,833 plus the land value of $700,000, giving a grand total of $3,844,833 as shown in Table 4.1.

4.18. *75% of Value*

A lender will generally provide a mortgage equal to about 75% of the business value of a proposed project. A mortgage is never equal to the total economic value of a new project (except for certain government-sponsored programs) because lenders want the developer to have a strong interest in making it a success. The 25% differential between the mortgage of 75% of value and the total value must be contributed by the developer and is called the *developer's equity.*

4.19. *Cost Approach*

From the lender's point of view, developers should have some hard cash equity in a project and should receive a reasonably attractive return on that cash equity. Therefore the lender wants to know the actual dollar cost of the project, from which the developer's investment can be determined. Then the percentage of return on the developer's cash equity investment can be computed. It is not necessary to furnish a builder's estimate, which shows

TABLE 4.3.   Construction Cost Summary for Picanos Villa Apartments, Urbane City, Anystate

Cash Needed	
Land	$ 700,000
Approximate construction cost	2,761,500
Loan brokerage at 2% (of $2,883,000)	57,660
Bank discount at 1% (of $2,883,000)	28,830
Construction interest, including discount	136,340
Closing costs, including title insurance	13,000
Taxes during construction	8,000
Total cash outlay	$3,705,330
Sources of Cash	
First mortgage	$2,883,000
3 months' cash flow, ¼ occupancy	47,056
Total mortgage and other income	$2,930,056
Estimated Cash Required	
(needed minus sources)	$775,274

a complete analysis of construction costs, but summaries should be presented somewhat as in Table 4.3.

4.20.   *"Construction Cost"*

The approximate construction cost is often called the *brick and mortar cost*. It includes site work and on-site and off-site development, in addition to the actual costs of materials and labor on the buildings themselves. Thus paving, sidewalks, landscaping, and all utilities are included, but not the builder's profit if the builder is also the developer. If the entire construction job is to be done by a third party, one who is not also the owner, then of course it is the third party's price to the developer that becomes the brick and mortar cost, and that price will include the builder's overhead and profit. The construction cost for feasibility purposes is often estimated on a unit cost in terms of dollars per square foot multiplied by the number of square feet of floor space in the project. For example in Picanos Villa Apartments, there are 78,900 square feet. If the unit cost is estimated at $35 per square foot, that multiplies out to 78,900 square feet × $35 per square foot = $2,761,500. Notice that in the example of Table 4.3 the construction cost amounts to $2,761,500 out of a total of $3,705,330, which is only 75% of the total cost. This percentage is typically in the neighborhood of 80% of the total.

4.21.   Replacement Cost

The total cash outlay is the true *replacement cost* of the project. All the expenditures listed under "Cash Needed" are unavoidable and hence necessary. In other words, if anyone in the world were to go out and build this project, and pay for everything necessary to complete the job, it would cost $3,705,330 (total cash outlay from Table 4.3). Everyone would have to pay the same interest, taxes, and so forth.

The fact that the replacement cost is less than the value by the income approach is of no particular concern, especially since $3,705,330 (replacement cost) is well within 10% of $3,844,833 (income value from Table 4.1). The two values are not expected to be the same. A lender usually will be quite satisfied if they are within 10% of each other. From the lender's viewpoint, replacement cost is less important than the economic worth of the business, because the lender must rely on the business income to repay the loan. Nevertheless it is expected that the replacement value of the property will be close to its income value as a business real estate venture. Furthermore, if the income value does not exceed the replacement cost, the project has problems and should be reconsidered.

4.22.   Financing Fees

The "loan brokerage" (Table 4.3) is discussed in greater detail in Chapter 5. The "bank discount" refers to the discount that is charged by the lender of the first mortgage. It is sometimes called an origination fee, and sometimes has other names. It could be omitted as a cost and deducted from the income calculation which follows the cost, but the first mortgage would nonetheless be recorded as $2,883,000 even though the discount had been deducted by the bank, so that the borrower had actually received only $2,883,000 − $28,830 = $2,854,170. Therefore, since the borrower has "spent" $28,830 and is obliged to pay back $2,883,000, the 1-point fee is shown under "Cash Needed," and the full mortgage is included under "Sources of Cash."

4.23.   Construction Interest

The construction interest can be estimated with a fair degree of accuracy if it is based on projected draw schedules, as will be shown in Chapter 10. However, for a preliminary analysis, a crude approximation will suffice. The amount of construction interest shown in Table 4.3 was based on several assumptions. It was assumed, first, that the job would take 12 months to complete, and then that all 12 of the construction draws would be equal. Actually, they will not be equal, but the assumption was made for the sake of simplicity. If a construction loan of $2,040,000 is assumed, each draw would be about $\frac{1}{12}$ of $2,040,000 or about $170,000. This is an approximation, but is satisfactory for the preliminary computation of construction interest. If the interest rate is 12% for 1 year, then it is $\frac{1}{12}$ of 12%, or about 1% per month. Using the methods detailed in Chapter 12 and 13, the total interest paid on the costruction loan is calculated as

$$\text{Total interest costs (TIC)} = \text{Accumulated principal and interest owed upon completion of construction minus principal only, or,}$$

$$\text{TIC} = F - nA$$

where

$$F = A(F/A,\ i,\ n)$$

$$\text{TIC} = A(F/A,\ i,\ n) - nA$$

$$= \$170{,}000\ (F/A,\ 1\%,\ 12) - 12 \times 170{,}000$$
$$12.682$$

$$= 2{,}155{,}940 - 2{,}040{,}000$$

$$\text{Total interest costs} = \$115{,}940$$

Now if it is assumed that the construction loan discount will be 1 point or 1% of $2,040,000 = $20,400, the total construction loan cost becomes $115,940 interest + $20,400 discount = $136,340.

4.24.   Sources of Cash—First Mortgage

It is hoped that the maximum "ceiling" amount of the mortgage will be funded, so the full mortgage amount of $2,883,000 is shown in Table 4.3 as flowing into the project. There is just no point in presenting a pessimistic picture to the bank, that will only tend to make the loan officer less enthusiastic about the request and more likely to turn it down completely. A bank is more likely to lend money to the developer who exudes confidence. Therefore the full amount of $2,883,000 is shown.

4.25.   Sources of Cash—Cash Flow

Some buildings will be completed and ready for occupancy before the rest. It is physically impossible to put the finishing touches on an entire building project at one instant of time. If a portion of it is completed some time before the rest of it, it is reasonable to assume that that portion can and will be rented before construction of all the buildings is finished. In the case being considered, it is assumed that an average of one-quarter of the project will be occupied for 3 months' time before the end of the year, producing a cash flow of $\frac{1}{4} \times \frac{3}{12} \times \$752,892 = \$47,056$, where \$752,892 is the gross potential rental income at 100% occupancy for a full year. Since the amount of \$47,056 is taken in before construction is complete, it is available to pay some of the costs incurred during construction, and is added to the cash obtained from the mortgage when computing the money available to pay for the project.

4.26. Income Feasibility

If a developer does not receive a sufficient cash return from project rentals, there may be a temptation to divert cash from accounts reserved for basic maintenance to meet the developer's cash needs. If, on the other hand, the ratio of cash return to investment is satisfactory, the developer is more likely to maintain the project to a high standard in order to preserve that satisfactory cash flow. The latter course of action will protect the investment of the lender, and the lender will more readily make a loan knowing that the borrower is going to receive a good return. The third phase of an income approach documentation is a feasibility analysis (Table 4.4), made on the basis of previous calculations.

TABLE 4.4. Feasibility Analysis: Picanos Villa Apartments, Urbane City, Anystate

Projected effective gross income (Table 4.1)		$710,190
Total operating expenses	$248,810	
Debt service at 12.639 constant rate $2,883,000 \times 0.12639$	364,382	
Total	$613,192	613,192
Net Cash Flow per Year		$ 96,998
Replacement cost (total cash outlay) (Table 4.3)	$3,705,330	
Total mortgage loan + other income (Table 4.3)	2,930,056	
Cash Equity	$ 775,274	

$$\frac{\text{Net cash flow per year}}{\text{Cash equity}} = \frac{96,998}{775,274} = 12.5\% \text{ Before-tax rate of return}$$

4.27.   Debt Service

Notice that up until this point no mention was made of payments on the mortgage. Now they are put into the computations as "debt service." The constant of 12.639 is the annual number of dollars required to repay $100 of borrowed mortgage money over a 25-year period with interest compounding at 1% per month (12% per year nominal). These constants are found in handy tables or can be calculated by methods shown in Chapters 12 and 13. Like the cap rate, the interest on the mortgage is set by the money market lenders and a current figure can be obtained from a mortgage lender at the time application is made.

4.28.   Cash Flow

The total cash actually paid out by a project during a year will include total operating expenses (in this case $248,810) plus payments of interest and principal and the mortgage ($364,382), for a grand total of $613,192 per year. Subtracting this figure from the projected income of $710,190 leaves $96,998 to go to the owner as cash profit each year, again based on present-day dollars. The annual cash profit is called the *net cash flow* per year.

4.29.   Before-Tax Rate of Return

Dividing the net cash flow of $96,998 by the cash equity of $775,274 yields a rate of return of 12.5%, before tax benefits and appreciation. The depreciable value of residential rental real estate currently can be depreciated over a 15-year period, yielding the same amount of depreciation each year for 15 years. For the example problem, this annual depreciation allowance (called "capital recovery" in the tax law) amounts to

$$\frac{\$3,005,330}{15 \text{ years}} = \$200,355/\text{year}$$

This amount is treated as a business expense and may be deducted every year for 15 years from taxable income before calculating the income tax due.

4.30.   Interest on the Mortgage

For a conventional fixed rate mortgage, the monthly mortgage payments normally remain constant throughout the 25-year life of the mortgage.

However, the fraction of that monthly payment designated to pay for interest starts out relatively high and gradually declines, while the remainder of the monthly payment which pays down the principal mortgage balance grows larger from payment to payment.

The interest part of the monthly payment is actually a form of rent paid for use of the mortgage money, and therefore is a business expense and is tax deductible. The principal portion of the monthly payment is simply repayment of a debt. Since loan money received by a borrower is not taxable income, neither is the repayment of the principal a tax deductible expense. So for tax purposes, the interest portion of the monthly payment must be separated from the principal portion. For a preliminary feasibility study a good approximation can be made by assuming the interest payment for the year is simply the mortgage balance at the beginning of the year times the annual interest rate. Therefore for the first year of the example project's life the amount of interest paid is calculated as:

Mortgage balance at BOY 1*	$2,883,000
Nominal annual interest rate	× 0.12
Approximate interest paid the first year	$ 345,960

4.31. Summing for the After-Tax Rate

Now the net taxable income can be found as the sum of gross income, plus auxiliary income, plus tax deductible depreciation, plus interest, plus operating expenses, as follows:

Effective gross income	+$710,190
Depreciation	− 200,355
Interest payments on mortgage	− 345,960
Operating expenses	− 248,810
Net taxable income	−$ 84,935

The resulting net taxable income is a negative amount, meaning that for income tax purposes this investment not only pays no income taxes but can shelter up to $84,935 of the owner's other taxable income. For an investor in the 35% tax bracket, $1 worth of tax shelter is worth $0.35 cash. That is, for $1 earned, the income tax paid would be $0.35 if the shelter were not available. So the $84,935 tax loss is worth $84,935 × 0.35 = $29,727

*BOY is used to designate *Beginning of Year*, while EOY designates *End of Year*.

in cash (assuming the investor/owners have that much additional taxable income that otherwise would be taxed at the 35% rate). This raises the rate of return calculated previously to

$$\frac{\text{Return}}{\text{Equity}} = \frac{96,998 + 29,727}{775,274} = 16.3\%$$

or almost one-third larger than the value calculated without tax considerations.

4.32. Appreciation

In addition, if the property is properly maintained and kept in good repair, and if it is well located, it will ordinarily *increase* in value at least over a significant part of its life. Thus, if the property increases in value by just 3% the first year, at EOY 1 it would be worth

$$1.03 \times \$3,705,330 = \$3,816,490$$

4.33. Capital Gains

If the project were sold at the end of the 1st year (EOY 1) there would be a capital gain and resulting capital gain taxes. The capital gain itself is defined as the difference between the depreciated straight-line book value and the actual selling price. In order to encourage capital investment, which is the key to greater productivity, most progressive governments give favorable treatment to capital gains. The income tax is paid at the ordinary tax rate (35% for our example) but only on 40% of the capital gain. This would be calculated as follows:

Total cash outlay cost new (original book value)	$3,705,330
Depreciation for the first year	200,355
Depreciated book value at EOY 1	$3,504,975
Appreciated resale price at EOY 1	3,816,490
Depreciated book value at EOY 1	3,504,975
Capital gain for first year	$ 311,515
Capital gain rate, 0.40 × ordinary rate =	
0.40 × 0.35 = 0.14	× 0.14
Capital gain tax if sold at EOY 1	$ 43,612

4.34. *After-Tax Rate of Return with Appreciation*

This raises the rate of return once again as shown by the cash flow calculations below. If the property were sold at EOY 1, the mortgage balance at EOY 1 must be paid off or assumed by the new buyer. The mortgage balance at EOY 1 is calculated by methods shown in Chapters 12 and 13 as $2,863,539. Therefore, the cash flow-out at EOY 1 is:

Appreciated resale price at EOY 1	+ $3,816,490
Mortgage paid off or assumed by buyer	− 2,863,539
Capital gain tax	− 43,612
Balance of cash to seller	+ $909,339

Now the rate of return can be calculated as the total cash out at the end of the year divided by total cash in at the beginning of the year, minus one (since the investment is being liquidated).

$$\frac{\text{Total out at EOY 1}}{\text{Total in at BOY 1}} = \frac{96,998 + 29,727 + 909,339}{775,274} - 1 = 33.6\%$$

Thus, if a 3% appreciation can be realized, the return rises to *33.6%* after taxes. Since tax laws are complex and subject to change, consultation with a competent attorney or accountant is necessary when determining the current tax rules applicable to any given situation.

4.35. *ROR Decreases as Owner's Equity Increases*

The project need not be sold at the EOY 1, and can be kept as long as desired. However, as the years pass, the owner's equity increases as the mortgage is paid down and as the value of the property appreciates. This increase in equity usually causes a decline in the rate of return, because the denominator usually grows at a faster rate than the numerator, causing the resulting percentage rate of return to decrease. Properties of this kind are typically resold or refinanced every 5 to 8 years.

4.36. *Conclusion*

After checking all market data, and all calculations, as well as all other aspects of the project, the developer/owner is satisfied that the project is feasible. Now comes the task of finding potential lenders and convincing them that the project is feasible. The next several chapters tell how this task is accomplished.

5

Sources of Financing:
How to Find a Friendly Lender

5.1. Needed—Good Funding Sources

If visions of development projects are to become realities, the successful builder/developer must find reliable sources of funding. The experienced developer customarily strives for maximum financial leverage, since the highest profit usually is obtained on projects using as much leveraging as possible. Leveraging means using as little as possible of one's own cash to control as much property as possible through the use of borrowed money.

Usually, two major loans are required to finance a project, although some lenders will combine these into one. The two loans are:

1. Short-term financing to finance the construction stage.
2. Long-term financing of a mortgage loan obtained upon completion of construction and used to finance the project over its normal operating life.

5.2. Short-Term and Permanent Financing

Short-term financing has several different forms and includes such loans as land purchase loans, land development loans, "front money," construction payroll loans, "gap" financing, and the construction loan itself. All of these terms will be explained later under their respective headings. But, in every case, short-term financing is provided by lenders only under the assurance that each short-term loan will eventually be repaid by the permanent mortgage loan. Hence the mortgage loan will be discussed first, because it constitutes the keystone of all construction financing. Without a firm commitment on the part of some reputable lender of long-term mortgage money, no short-term loans can be obtained, and the project may not be built, unless of course the owners are in a position to pay all of the bills with their own money.

5.3. The Mortgage Loan

There are several sources from which a developer can obtain permanent financing for a project. The source to approach depends on the size and nature of the project. In some instances a particular lending institution can handle any type and size of real estate venture. But there are some financial institutions, such as insurance companies, that will not lend *less* than several hundred thousand dollars on any one project. On the other hand certain small savings and loan banks cannot lend *more* than a limited amount on any one job, being restricted by law because of their status as a financial institution. However, even when a certain small bank is incapable of handling a larger loan, it can join forces temporarily with several other such banks, and each will participate in a given loan, thus forming a joint venture in financing. In that case the originating bank handles the details of getting together the total sum requested by the developer. The developer usually has no contact at all with the other participating banks since all of the joint loan arrangements with the other banks are worked out by the originating bank.

5.4. Lending Institutions

Although all of the various institutions that lend long-term money for real estate projects at times may be referred to as banks, there are actually five types of lending institutions:

1. Savings and loan associations
2. Insurance companies
3. Quasi-governmental corporations
4. Pension funds and other trusts
5. Real estate investment trusts (REITs) or other specialized private investors

5.5.   *Mortgage Brokers and Mortgage Bankers*

Each of the five financial institutions enumerated above offers certain advantages and disadvantages when compared with the other four, and within each type there are a number of individual firms doing business. To avoid the guesswork of trying to choose the specific institution or firm that may have funds for the particular project being proposed, and to save considerable time, it is usually advisable to employ the services of a mortgage broker or a mortgage banker. In fact, it is sometimes impossible to approach certain lending companies without going through a mortgage broker or banker.

5.6.   *A Mortgage Broker*

Mortgage brokers do not provide their own funds for any project. They are professionals whose job it is to be middlemen between borrower and lender, collecting a fee for matching would-be borrowers with appropriate lenders. The broker's job is to arrange the loan and that is all. After the loan is closed the broker has no more interest in it. The broker's fee usually is about 1 to 2% of the loan amount. In a number of states mortgage brokers are licensed by the state after passing a written examination. Many real estate brokers have also become licensed as mortgage brokers, since in the course of selling real estate they frequently encounter would-be purchasers in need of mortgage financing.

5.7.   *A Mortgage Banker*

In addition to providing the services of a mortgage broker, mortgage bankers also may lend their own money for the mortgage loan and then sell the mortgage to a third-party lender. The mortgage banker usually has one or more large lending institutions as clients. These lenders frequently are headquartered in capital surplus areas where the supply of money is more than adequate to meet local demands. Therefore, in these capital surplus areas there is more money to lend than is needed for worthwhile projects, and, due to lack of local demand, the local interest rates are lower. On the other hand, in capital deficient areas, usually a larger volume of new construction must be financed to meet the demands of rapid growth. Since there is a greater unsatisfied demand for money to borrow, the interest rates are higher. The mortgage banker's job is to locate lenders with money in capital surplus areas and channel the surplus funds to borrowers in need of money in capital deficient areas. The mortgage banker finds qualified borrowers who are reliable security risks and who want to borrow money secured by mortgages on good property.

5.8. *Qualifying the Borrower*

Qualifying the prospective borrower involves comparing the characteristics of the prospective borrower to the criteria established by the client lending institution. Qualifying basically involves checking into assurances that the borrower possesses the three fundamental "C"s of a good borrower:

1. *Character.* The borrower has the desire to repay and has habitually repaid all obligations in the past. If a borrower is reluctant to repay his just debts, the high cost of forcing collection can make loans at normal interest rates unprofitable to any lender.

2. *Competence.* The borrower is competent to put the loaned funds to a profitable use. A borrower making a good profit is much more likely to repay on time than one who is incompetent and loses money. Competence insures that the money loaned won't be squandered or wasted due to incompetence, and thus decreases the risk to the lender.

3. *Collateral.* The borrower has sufficient resources to guarantee the repayment of the loan. The lender needs assurance that the mortgage loan money will be invested in the construction or purchase of a sound, well-built structure in a good location that will encourage preservation and appreciation of value and will protect against rapid deterioration or depreciation. This also assures the lender that if foreclosures should become necessary, the borrower's equity in the property is sufficient to adequately cover the defaulted loan.

5.9. *Servicing the Loan*

Once the borrower is found and qualified, and the loan is closed, the mortgage banker, like the broker, is entitled to a fee for this portion of the work. Unlike the broker, however, the mortgage banking firm will frequently lend their own funds to the borrower and then retrieve these funds by selling the mortgage to the client lending institution located in a distant city. Usually along with the sale of the mortgage goes an agreement by the mortgage banker to service the mortgage, including collection of principal and interest payments, and if appropriate, collection of escrow funds, with the accompanying duties of disbursement of taxes and insurance. Also the mortgage banker may conduct periodic inspections to insure proper maintenance of the property, and, should they become necessary, conduct collection and foreclosure proceedings. For servicing the mortgage the mortgage banker typically receives a monthly fee of one-half of 1% of the remaining balance due on the mortgage. So complete are these servicing functions that the borrower may never even realize that the ownership of the mortgage has left the hands of the mortgage banker.

5.10. Transferability of Mortgages

Unless the mortgage stipulates to the contrary, lenders are free to sell the mortgage to whomsoever they desire, and borrowers are free to sell the property with the new buyers assuming the obligations of the old mortgage. However, due to the fear that rising interest rates might depress the resale value of the mortgage, most new mortgages issued since the latter 1970s have contained a "due-on-sale" clause prohibiting new buyers from assuming the old mortgage without permission from the lender. This permission frequently is granted only with an upward adjustment in interest rates, especially if interest rates have risen since the mortgage loan was originally closed.

5.11. Channeling Funds to Capital-Short Areas

The mortgage banking industry provides a valuable service to both borrowers and lenders. Areas of rapid growth such as Florida, Texas, Arizona, and California usually do not generate sufficient local funds to support their own rapid growth. That is, the citizens, firms, and corporations of the local growth area do not deposit enough money into lending institutions within the area (such as Savings and Loan Associations, retirement trust funds, insurance companies, etc.) to supply the demand of borrowers for loans to fund the rapid growth. On the other hand, some of the older, more established areas of the country (such as some northeastern cities) generate more savings, insurance premiums, and retirement funds than they can handily invest in the area. The mortgage banking industry provides the channels by which the funds can be routed from areas of capital surplus to areas that are short of capital. If borrowers in Orlando, Florida need a few million dollars for sound well-conceived projects, the mortgage bankers in that area may well locate surplus funds in Boston, Massachusetts and channel them quickly to service the need.

5.12. Broker/Banker Expertise

Both the mortgage broker and mortgage banker typically can look to several sources of loan funds and by necessity are in constant touch with the money market. Typically each lender will have a slightly different format for providing the information required for approval of a loan application, and each lender may turn down a request if wrong or insufficient data is presented, or if the data is not presented in a form approved by the institution. The mortgage broker/bankers serve as experienced guides through this paperwork maze as they are familiar with each lender's preferences and idiosyncrasies and not only can select an appropriate lender for any given project but can greatly expedite the request for funds directed toward that lender.

5.13. Professional Fairness Required

The nature of their work requires that the broker/bankers must be fair to both the lender and the borrower. If, for example, they encourage the bank to put too much money into a given project, that project may find itself in grave financial difficulty. The mortgage payments may be so large that other obligations cannot be met, and the owner, unable to pay all the bills, may abandon the project, and turn it back to the lender. Lending institutions do not want to be landlords—their expertise and greatest profit potential lies in the money-lending business, not in the real estate business. Therefore, if the broker/banker is overly optimistic and presents figures that are not realistic, the lenders will soon quit trusting him and will refuse to finance projects coming from his office. Then if the broker/bankers sources of funds dry up, they will be out of business. On the other hand, the broker/bankers cannot be too pessimistic either. If a prospective borrower can show a high probability of success providing the project can get proper financing, the broker/bankers should exert every effort to obtain as high an amount of money for their client as reasonably can be justified rather than some lesser amount. If the broker/bankers consistently fall short of finding enough money to adequately finance their projects, their customers will soon go elsewhere and, again, their business will dry up.

5.14. Importance of the Loan Application

Even after acquiring the best available land parcel under good short-term financing conditions, the success or failure of a real estate venture still hinges upon securing an adequate permanent mortgage loan at a reasonable interest rate and with a long pay-out period. The amount of the permanent loan on any given project can be increased by a well-planned and well-documented presentation by the borrower. Therefore, the application for the loan becomes extremely important in obtaining this project keystone—the permanent mortgage loan. A well-organized loan application, complete, logical, and neatly presented leaves the impression that the loan applicant is a skilled professional. The lender may even be inclined to lend more money than usual assuming that the entire project is as well planned as the presentation, and will have less than ordinary risk. The form of presentation influences a lender both directly as well as indirectly. A skilled presentation gives the lender a subconscious assurance that the loan applicant is a competent, reliable, and trustworthy professional. Naturally a competent professional can be expected to do a much better job than the average borrower and doubtless will manage the business of developing the project well and will make a well-deserved profit, thus reducing the risk to the lender by assuring adequate income to repay the loan.

5.15. Reducing the Owner's Equity

The applicant's bonus for an outstanding presentation might amount to an increase of as much as 5 to 10% on the amount of funds loaned. On a $1,000,000 project a thorough and well-documented proposal could mean a loan of $50,000 to $100,000 more than that obtained by the ordinary loan application. If the developer can obtain a loan that is larger than ordinary by $50,000 to $100,000, the developer will not have to invest that $50,000 to $100,000 out of the developer's own precious liquid capital funds. When the developer places capital in a project this capital is called owner's *equity* capital in the project. The larger the loan, the less owner's equity is required. If the loan is large enough, it is sometimes possible to eliminate the owner's cash equity requirement completely. However, most lenders prefer a substantial owner's cash equity to insure that the project is worth more than the debt owed on it in the unlikely event the lenders have to foreclose on the project.

5.16. How Much Is the Loan Value?

On a single-family home the loan value would be based on an estimated sales price, or at least on an appraisal of the market value made by a lending officer of the bank. The size of the house, its amenities and materials of construction, the size of the lot, and the quality of the neighborhood are all taken into consideration in arriving at an appraisal of the value of the completed building, and the mortgage loan amount equals a certain percentage of that value. But on a commercial venture the loan is typically based on the economic worth of the project, that is, not on the value of the buildings and land, but on the value of the project as a profit producing business.

For a real estate development such as an apartment, a motel, or a shopping center, the buildings as buildings, and the land as land, are not worth much to the bank. The bank does not want to own buildings or land. The whole project has a value only when it is considered as a profitable business. The business will produce money, because according to the projections, the cash flowing in will exceed the cash flowing out. After all the money is collected from rentals and all expenses have been paid, there will be cash left over to make regular repayments on the loan and to give the developer a reasonable profit. It is only the money that flows out from the business that can pay off the loan—an empty building cannot generate cash flow simply as an empty shell.

The loan, then, is to be paid back from the proceeds of a business venture, and the bank will lend money when it can see how it is going to be paid back. The bank will not lend money solely because the landscaping for the project is beautiful and the design of the buildings is excellent.

5.17. Preliminary Steps for a Loan Application

Before approaching one of the lending institutions with a loan application, the broker/banker will check the developer's estimates of cost and income against the current market, visit the site of the proposed project, and develop an opinion of the venture. To aid the lender in formulating an opinion the developer should present two items, (1) a rendering (sketch) of the completed project, and, (2) complete estimates of costs and income as discussed in Chapter 4. These two items are often referred to in good humor as "a pretty picture and a price." The sketch should include preliminary designs by an architect which show the layout in sketch form, the appearance of the various buildings, and an overall view of the project. An important part of an application for a loan is the financial statement of the developer, or for a partnership, a financial statement of each one of the partners.

5.18. What's Included with a Loan Application?

An application for a loan on a smaller real estate project should include *plans, specifications,* and usually a *financial statement* of the borrower. The bank usually furnishes a standard form for the borrower's financial statement. The architectural plans should be fairly complete. In the case of a single-family residence or duplex, the law in most states permits builders to draw the plans if they care to. However, plans for most other projects must be drawn by an architect. The term "plans" includes specifications. The specifications on dwellings and certain other projects can be rather sketchy since building practices generally are well constrained by building codes and acceptable standard practice. However, on commercial buildings, such as an office building, for example, specifications usually need to be more precise and extensive, even if the building itself is not very large.

In addition a surveyor may be needed to obtain field measurements and draw the plot plan showing the legal description of the land, the lot and block number, name of subdivision, and other required data.

On small projects loan applications are usually made to local banks located in the same geographic area as the project. Therefore the local bank will already be familiar with the geography, demography, and other factors affecting the neighborhood of the project. The local banker may even be acquainted with the borrower, and so will have a far greater sense of "feeling" for the project than some out-of-town lender far removed from the project. Because of this familiarity the local bank can make a decision to accept or reject the loan much more rapidly. Usually, the bank's loan committee meets once or twice each week to decide on all applications, so a well-documented request for a loan might receive a response in as short a time as one week.

5.19.   For Smaller Loans Apply Directly to the Bank

It is not always necessary to work with a mortgage broker or banker. For a mortgage on a small project, such as a house, a small apartment building, or a small office building, it may be advantageous to go directly to a savings and loan bank and present the request to the mortgage loan officer in person. The advantage of this procedure over going through a mortgage broker lies in a saving in time and money. An application for a loan made through a mortgage broker may take as long as 30 days for approval or disapproval.

The mortgage loan officer of a savings bank, as an employee of the bank, not only does the same work without a broker's fee, but the results are obtained somewhat faster too. The loan officer can determine, usually within a week or two, whether the bank's loan committee has approved the application. The loan officer does not have the authority to make a loan in the name of the bank, but can and will advise the borrower on the probabilities of the loan being approved. Similarly, the loan officer can provide current information on the going rate of interest, amount and number of repayments, the various fees that will have to be paid, prepayment privileges, and other provisions of the loan contract.

5.20.   Persistence up to a Point

If the first broker/banker contacted rejects the project, then contact at least one more. If the second one does not approve the project, there may be some serious flaws in the proposal, and the whole concept should be carefully reviewed.

5.21.   Closing Costs

For a comparatively small project, the closing costs on a real estate mortgage loan include:

Discount points, origination fee, credit report
Legal fees
Recording fees and documentary stamps
Abstract, title insurance, attorneys title opinion, survey, photos

For larger real estate ventures other items are included.

5.22. *Discount Points, Origination Fee, Credit Report*

These are one-time fees charged at the time the mortgage money is loaned in order to cover the lender's costs incurred in qualifying the borrower and putting the loan on the books. Since they occur at the front end of the pay-back period, they are sometimes called front-end charges. Discount points is the percentage of the value of the mortgage added to the other front-end charges. For instance, 3 points on a $100,000 mortgage amounts to 3% of $100,000 or $0.03 \times \$100,000 = \$3000$. The discount points are usually added to the amount repaid (e.g., borrow $100,000 and repay $103,000), but may by mutual agreement be subtracted from the amount borrowed (e.g., borrow $100,000 less $3000 = $97,000 and repay $100,000).

Discount points are frequently considered as a trade-off for higher interest rates on the body of the mortgage (or a payment of interest in advance). One point of front-end cost is about equivalent to $1/8\%$ interest on the mortgage for long-term mortgages. Thus if the market interest rate for mortgages is $13\frac{7}{8}\%$, then a certain bank may offer to lend at an equivalent $13\frac{3}{8}\%$ plus 4 points (they are equivalent since $13\frac{3}{8}\% + 4$ points $\times$ $1/8$ point/$1\% = 13\frac{7}{8}\%$). For a borrower intending to pay out the mortgage over its full term, there is no significant difference between $13\frac{3}{8}\%$ plus 4 points and $13\frac{7}{8}\%$ and no points. The advantage to the lender occurs when, if interest rates should drop, the borrower is not nearly so eager to refinance the loan since part of the interest essentially has been paid in advance through payment of the points and is nonrefundable. If the borrower does refinance, the lender retains the points already charged as partial compensation for the extra cost of closing out the loan and the turnaround costs involved in finding a new borrower qualified to use the money. To additionally compensate the lender for the costs of early repayment or re-financing of the loan, lenders will often insert penalty clauses for early repayment. A typical penalty clause may include "In the event the borrower shall prepay this loan before the expiration of the term of the loan, the borrower shall obtain a refund of the unearned portion of the interest charge computed under the Rule of 78s method." This Rule of 78s results in an unbalancing of the interest charges so that the actual interest rate starts out higher than the nominal average rate and declines slightly each month until the actual interest rate ends up lower than the nominal average during the last half of the mortgage life.

Another typical clause states "If the prepayment of the original principal amount of the loan exceeds 20% in the first 3 years of the loan, 2% of the excess will be charged as a penalty." For example on a $100,000 loan, if the borrower for some reason wants to prepay a lump sum of $45,000 before the end of the first 3 years, he is permitted to repay the first $20,000 without penalty, since that is included in the first 20% of $100,000. But the borrower must pay a 2% penalty on the remaining $25,000 (penalty for prepayment is $25,000 × 0.02 = $500).

5.23. Legal Fees

Since rights to real estate can become involved with a lot of legal technicalities, the lending institutions insist on having their attorneys review all aspects of the loan. The borrower is required to pay the legal fees charged by the bank's attorney. In addition the borrower may want his own attorney to assist with the transaction and approve all papers. The borrower also pays those fees, in addition to the bank's attorney's fees. Normally the bank will not accept work done by the borrower's attorney, nor will it permit substitutions of title insurance for a part of the legal work, nor will it accept an opinion of title to the land by anyone else. The opinion of title, together with all other necessary documents, must be prepared by the bank's attorney, and the borrower must pay the fee that the attorney sets. Due to competition between banks for the borrower's business, the fees usually do not reach exhorbitant levels, and the paperwork usually proceeds quite smoothly using the bank's experienced attorney. Furthermore, if both parties consent, it is usually fair and ethical for one attorney to draw up an agreement between two clients that are in accord. Although the borrower and the bank share the services of the same attorney, the borrower pays the entire legal fee. The various papers that the attorney draws up often are referred to as "legal instruments."

5.24. The Note and the Mortgage

The borrower will be required to sign two legal instruments, a *note* and a *mortgage*. The note is evidence of indebtedness and constitutes a personal pledge to pay back the amount borrowed. A note is transferable and can be given or sold from one party to another. A mortgage, on the other hand, is evidence of security. The mortgage is recorded in an appropriate book at the county courthouse of the county in which the property lies. Anyone concerned with the property then has the opportunity to check up on any liens or transactions involving the property. Thus prospective buyers, lenders, or anyone else can obtain a complete listing of debts or encumberances of any kind against any property in the county and learn whether or not these encumberances have been satisfied. The mortgage shows that if the borrower who issued the note accompanying the mortgage cannot pay when the payment falls due, the mortgage lender has a lien on the property as security. The property is a tangible item, normally of equal or greater value, which can be sold to repay the loan in full. A default on the mortgage usually is considered as a breach of the mortgage agreement. Not making mortgage payments on time is just one way of defaulting on the mortgage agreement. The borrower also may have agreed to carry insurance on the property, or to carry a termite contract, or to keep the property in good repair. A breach of any of these or any other provisions could constitute a default. In case of default, the lender files for foreclosure in the county in which the property is located, and if the court does not find to the contrary, a foreclosure judgment is obtained. The sheriff then takes possession of the property in the name of the court and, after public notice, auctions it off at public auction to the highest bidder. Anyone can bid including the owner and the lender. The proceeds of the sale are used to pay off the balance of the mortgage note. Any excess cash from the sale, after the note and legal fees and any other liens are paid, goes to the original owner. If the foreclosure sale does not bring enough money to pay off the note, the borrower still owes the balance.

5.25. Recording Fees and Documentary Stamps

When a change in ownership interest in real estate (such as a sale or mortgage) is placed on the public record at the courthouse, a recording fee is charged. These recording fees include a nominal charge by the County Clerk or other appropriate county officer for the actual recording of the mortgage, plus a tax of around ½ of one percent of the transaction price charged by the state government on such transactions.

5.26. Abstract, Attorney's Title Opinion, Title Insurance, and Survey

The abstract, the attorney's title opinion, title insurance, and the survey of the property are all designed to assure the lender that the property is good and proper security for the loan. In order to accept the property as security, the lender must know that the property does legally exist, and is located where it is supposed to be (e.g., the buildings don't encroach on adjoining property), that the borrower is the legal owner, and that there are no undeclared or unknown encumberances, liens, or other defects in the title. The abstract is a history of the property composed of a collection of copies of all the deeds, mortgages, and other conveyances and encumberances by previous owners of the property and the transactions affecting the property. This collection usually costs several hundred dollars worth of attorney's and clerk's time to compile. After the abstract is compiled an attorney still has to examine it and render an opinion as to the validity of each transaction and the resulting title. Many times a title insurance policy will be substituted for the lengthy abstract. The title insurance company usually insures the title against a limited list of title defects, and then further limits its liability to an amount not in excess of the purchase price of the property at the time the policy is issued. In the unlikely event a flaw is found in the title after a number of years of ownership, the title insurance company could either refuse to pay if the flaw were of a type excluded from coverage, or pay the value of the property at the time the policy was originally purchased, likely much less than the value of the property at the later date of discovery of the title flaw.

5.27. Closing Costs

All of the costs that are incurred to qualify the borrower for loan, as well as to qualify the property as adequate security for the loan, are charged to the borrower as either application fees, paid at the time of application for the loan, or paid as closing costs at the time the loan is actually granted. The borrower's closing costs, including application fees, typically range from 1½ to 4% of the mortgage loan amount. For example, on a $100,000 loan the closing costs could range from about $1500 to $4000. Each lender may vary from others in its requirements, criteria, and charges, so prospective borrowers sometimes can reduce their borrowing costs significantly by shopping around and comparing rates among lenders.

6

Going for the Commitment

6.1. The Commitment

At this stage the application for a mortgage loan has been submitted to a mortgage broker, and the developer is waiting for a lender to promise, providing that certain conditions are fulfilled, to lend the mortgage money. This promise by the lender is called a "commitment." The mortgage broker has carefully reviewed the application, submitted it in a form acceptable to a particular bank or other lender, and has forwarded it to that institution, perhaps even appearing in person to present the case. If the lender's criteria are met, and money is available, then after careful consideration the loan committee of the lender probably will approve the application. However, no money will change hands yet. Instead, the lender will issue a formal document which commits it to lending a specified sum of money to a specified borrower, provided the borrower fulfills the several conditions set forth in the statement. The document is called a *commitment* and is sent from the lender to its agent, the mortgage broker, who then puts it into his own form and transmits it to his client, the borrower. Alternatively, as previously explained, the commitment can come directly from a Savings & Loan Association bypassing the mortgage broker. There are also times when a mortgage broker will deal with a Savings & Loan Association.

6.2. Commitment Form Explained

The commitment letter shown as Figure 6.1 will now be discussed in detail. Figure 6.1 shows a form that is used by one particular Savings & Loan Association. It is not universal. Each company devises its own form, but this form is representative and contains the information and data normally

Empire of America FSA

(formerly First Federal Mid-Florida)
P.O. Drawer FS, Gainesville, Florida 32602

March 29, 19

RE: Green Grove Apartments, a 48-unit
     apartment complex located in
     Orlando, Florida

Maple Hill, Ltd.
2830 West Highway Drive
Orlando, Florida

Gentlemen:

We are pleased to advise that this Bank has approved your application for a first mortgage term loan on the above-described apartment complex.

This financing shall be represented by a note and mortgage in our usual form and containing the following terms:

Mortgagor:	Maple Hill, Ltd.
Obligor:	Maple Hill, Ltd.
Guarantor:	Robert Metcalf Harry Metcalf, M.D.
Amount of Mortgage:	$1,200,000
Term of Mortgage:	15 years (30 year amortization)
Initial Interest Rate:	11.50% per annum
Adjustments:	The interest rate is to be adjusted quarterly to 1% above the Chase Manhattan Bank prime rate.
Prepayment Privilege:	Prepayment shall be permitted during the first year of the term of this mortgage loan with a 7% prepayment charge during the first year, said prepayment charge declining 1% per year during the term of this loan or any renewal or extension thereof. The principal balance may be paid in full, without penalty, if the loan is called at maturity.
Escrow:	A tax and insurance escrow will be established at the time of closing and monthly payments will be made thereto.

FIGURE 6.1. Commitment letter.

Empire of America FSA

(formerly First Federal Mid-Florida)
P.O. Drawer FS, Gainesville, Florida 32602

Subject to the following terms and conditions:

1. All costs in connection with this loan, including the Browns and Associates legal fees and disbursements of our attorneys, cost of mortgagee title insurance premium, survey, appraisal cost, recording charges and filing fees, any state mortgage tax and revenue and documentary stamps, if subsequently found to be required, shall be paid by you. All such costs which we disburse and our attorneys legal fees and disbursements must be paid not later than the day of closing.

2. The loan documents will provide that a 4% late charge will be imposed on any payment not received by its due date (or by the first business day thereafter if the due day is not a business day for this Association), and that the interest rate on the loan shall be increased by one percent (1%) per annum from the time we have a right to accelerate the indebtedness upon the occurrence of a default under the loan documents until the time such default is cured.

3. All documents, title questions, designation as to applicable law and all other legal matters relating to or arising out of this loan will be subject in all respects to the approval of our attorneys.

We shall require that you furnish to us a mortgagee title insurance policy in the face amount of this loan issued by a company satisfactory to us or A.L.T.A. form certifying the marketability of title to the premises being mortgaged, vesting of the same in the mortgagor and freedom from mechanics' liens and comparable claims. All exceptions in said policy are to be subject to the approval of our attorneys.

We also require that you furnish us at least 20 days prior to closing with a current survey satisfactory to our attorneys and conforming to the "Survey Requirements" annexed at the end of this commitment letter.

4. You represent and agree that the premises being mortgaged will be insured (naming this Association as first mortgagee) against loss by fire and such other hazards (including flood insurance if the subject premises are within a flood risk area as designated pursuant to the Flood Disaster Protection Act of 1973) as may reasonably be required by us in an amount and with a company or companies satisfactory to us and also that you will furnish extended coverage. One or more policies, binders or certificates of insurance evidencing such coverage must be delivered to us before closing.

5. You represent and agree that the improvements to be constructed on the premises will comply in all respects with local zoning and use requirements and you shall furnish us prior to closing with a certificate of occupancy or equivalent certification issued by the appropriate local governmental agency, evidencing compliance with such zoning, use and occupancy requirements.

6. You shall furnish us with a fully documented appraisal of the premises and the improvements thereon, which appraisal shall be satisfactory to us in all respects and shall be made by an M.A.I. or SREA appraiser, who must be approved by this Bank prior to your ordering said appraisal.

Empire of America FSA

(formerly First Federal Mid-Florida)
P.O. Drawer FS, Gainesville, Florida 32602

Maple Hill, Ltd., cont'd.
March 29, 19
Page Three

7. References in this commitment to "approval" and "satisfactory" shall not be interpreted as justifying arbitrary rejection, but shall connote a reasonable application of judgment taking into consideration this Association's custom concerning major real estate transactions.

8. The terms of this commitment shall survive the execution and delivery of the note, mortgage, and all other documents herein referred to. This commitment is not assignable.

Sincerely,

Robert E. Cameron

Robert E. Cameron
First Vice President

provided in most commitments. Once the commitment is signed by the borrower it becomes a binding contract obligating the borrower to construct the project according to the plans submitted and according to any stipulations contained in the commitment. The lender is similarly obliged to furnish the stipulated sum of money at the stated interest rate for the stated period of time.

6.3. Letter Form

This commitment is in the form of a typed letter. In the past, commitments often consisted of printed forms with blank spaces filled in, but today it is common practice to use the customized letter style since terms and conditions change so rapidly in the money market.

In this case, the lender is the Empire of America Federal Savings Association. The commitment is dated for two reasons. First, any business transaction or any document related to business should always be dated to prove priority in case of any future modifications to the agreement. Second, a commitment is a timely instrument that is valid only for a specified period of time, generally 30 to 90 days, and the borrower must accept, reject, or ask for modification before the stipulated time period expires.

This letter form is considered as a commitment since a particular project is named and described. The "Green Grove Apartments" are to be a 48-unit complex located in Orlando, Florida.

The borrower will be Maple Hill, Limited. The word "Limited" in the title means that the company is a limited partnership, a form of business described previously in this book. The address of the limited partnership is: 2830 West Highway Drive, Orlando, Florida. Ordinarily the property location is given by a full legal description, involving metes and bounds, or subdivision blocks and lots and address. The location has been edited out of this example. The location designated is the location inspected and approved by the lender, and the lender requires that the project be built on that approved site. Although no drawings are referred to in the commitment, drawings are usually submitted with the application for the commitment. Therefore, when the commitment states that the project must be a 48-unit apartment complex it infers that it will be constructed in substantial accordance with the plans previously submitted.

Since the commitment obligates the borrower to build the project according to the plans submitted it is obvious that the preliminary plans submitted with the application should not be just rough sketches, but should be thoroughly thought out and carefully presented. For example, an architect once sketched a preliminary design depicting a beautiful oak tree in front of an office building. The bank later insisted that this oak tree be planted by the builder, claiming that one of the inducements for making the loan was the appearance of that tree in front of the building. The mortgage is committed for a specific project, and the project must be built as it was submitted for approval.

6.4.   Mortgagor and Obligor

The mortgagor is to be Maple Hill, Limited. This means that the one who owns the property on which this mortgage is placed, and who signs the mortgage as mortgagor is the limited partnership of Maple Hill, Limited. The one who borrows the money is also Maple Hill, Limited and is herein called the obligor. The mortgagor and obligor could be two different companies or two different individuals if one owned the property and the other borrowed the money using the property as mortgage collateral.

6.5.   First Mortgages and Other Mortgages

This application was made for a commitment for a first mortgage loan. It is possible to have any number of mortgages on the same property. There could be a first mortgage, a second mortgage, a third mortgage, and so on, or there can be a first mortgage and two second mortgages. The difference between a first mortgage and a second mortgage is usually a matter of timing sequence, but not always. The first mortgage is not always chronologically the first mortgage to be placed on the property. Rather, the first mortgage designates a ranking of importance. The second mortgage is subordinate, or junior, to the first mortgage, and the first mortgage takes precedence over the second mortgage. If a partial monthly mortgage payment is made not adequate to cover both the first and second mortgage, then payments are made first to the first mortgage, then to the second mortgage, and then to any other bills that have to be paid. If there is a default in payments to the first mortgage, the first mortgage lender files in the courts for foreclosure. At the direction of the court, the county sheriff then takes possession of the property, and the property is sold at public auction. The proceeds from the sale are used to pay off the first lien, which is usually the first mortgage if no tax liens are present. Any cash remaining after payment of the first mortgage is used to pay off the second mortgage and any other subordinate liens on the property. If any money still remains after all claims against the property are paid, it goes to the former owner.

In all cases of multiple mortgages on one piece of property, the mortgages other than the first mortgage are termed *subordinated*. The lender of the first mortgage usually requires that all other mortgages be subordinated to the first mortgage position, and that they be so noted in the county records so that the public will be on notice as to whose mortgage is first. The term "subordinated" means that these mortgages are inferior in priority and junior to the claims of the first mortgage. A third mortgage is subordinate to the second mortgage, which in turn is subordinate to the first mortgage.

Since the numbering of the mortgages indicates priority of lien rather than timing it is possible to obtain a second mortgage before getting a first mortgage. Such a case occurs when the lender of the second mortgage agrees to a subordination to a future first mortgage. For example, when land is purchased the land seller could agree to subordinate the mortgage on the land to a future mortgage on buildings yet to be built. The mortgage on the land then is recorded immediately in the county records as a second mortgage, subordinated to a future first mortgage. If no first mortgage is ever placed on the property, then the second mortgage, being the only debt on the land, has first priority just as a first mortgage would have.

6.6. Guarantor

The guarantors are Robert Metcalf and Harry Metcalf. This indicates that the mortgage payments are guaranteed by the two individuals Robert and Harry Metcalf, and that if the project gets into financial trouble these individuals will make the mortgage payments out of their own resources. The developers usually prefer to have the project stand on its own, leaving the individual borrowers with no personal liability. Then if the project does get into financial difficulty the bank cannot look to the personal assets of the borrowers to make up any losses on the project. However, the banks recognize that it is in their best interests to try to obtain as much security as possible to guarantee repayment of their loans. Therefore they may charge higher interest rates or refuse the loan altogether if the extra security is not forthcoming.

6.7. *Terms of the Mortgage*

The next item in the commitment shows that the lender is willing to lend $1,200,000 on this project. The 48-unit project will probably be appraised at about $1,700,000 and thus the lender is committing for a mortgage amounting to only about 71% of the cost of construction. Hence the developer/borrower has an equity position of the difference, which is about 29%. This is a comparatively large equity position and because the developer/borrower is putting so much of his own money into the project, the lender is giving comparatively favorable terms here. The initial rate of 11½% per year is below the current market rate at the time of the commitment.

The term of the mortgage is 15 years, which means that there will be (12 × 15 =) 180 monthly mortgage payments over a 15-year period. However, there is an added phrase: "(30-year amortization)," which means that payments are made at a monthly rate that would pay out (amortize) the mortgage over a period of 30 years instead of 15 years. At the end of 15 years the owners must either pay the mortgage balance remaining (as a lump-sum balloon payment), or refinance the mortgage. This is a form of balloon note which will be explained later. However, at least during the first three months of the mortgage (since the rate may be adjusted quarterly) the owners will pay monthly mortgage payments at 11½% per year on a payment schedule that would pay off the mortgage in 30 years. The interest rate of 11½% is tied to the prime rate and is subject to change every three months (quarterly) if the prime rate changes. If there is no change in the prime interest rate, then these mortgage payments would remain constant for 15 years. At the end of 15 years (one-half of the amortization period), the owner will have paid far less than half of the principal. In fact the remaining principal balance is $1,017,258. The bank would than have the option of demanding that this mortgage balance be paid in full immediately or of offering new financing to the borrower on new current mortgage terms. If the bank were to demand the entire balance due, the borrower either goes to another lender to borrow the amount required as a new mortgage, or the borrower petitions the original lender for new terms and an extension on the old mortgage.

6.8. *Interest Rate*

Although the initial interest rate is listed as 11½% per year, the adjustment indicates that the rate is subject to change quarterly so as to remain at 1% above the prime rate of the Chase Manhattan Bank. The prime rate is the rate of interest the bank charges its most favored customers. This rate fluctuates from time to time and not long ago was as high as 20%. Conversely, within 5 years of the peak prime rate it was as low as 5%. The bank's main concern is that the interest income it receives for money loaned out should adequately reflect the cost the bank has to pay for borrowing the money from its investors and others. If inflation does occur in future years the borrower usually can afford to pay more interest since the value of the mortgaged property is increasing in proportion to the rise in inflation, and higher rents will usually offset the higher mortgage payments.

6.9.   The Constant

In the past, the standard fixed rate form of mortgage required a constant monthly payment that did not change over the life of the mortgage. The interest rate was fixed as was the length of the term of the mortgage. Tables were provided that showed the percentage of the loan to be paid each year. For example, for a loan at 11½% and 30 years, the constant would be 11.88. This indicates that each year the total of principal plus interest paid would be 0.1188 times the amount of the loan. For a loan of $1,200,000 the annual amount paid would be about 0.1188 × $1,200,000 = $142,560 per year, or monthly about $142,560/12 = $11,880 per month. The constant can still be useful in finding the readjusted monthly payments, given the new interest rate, the remaining time, and the principal balance.

6.10.   Prepayment Privilege

Because interest rates might decline in the future, the owner could find it advantageous to pay off or refinance the existing mortgage at some future date. Of course the bank makes its profit by charging interest over the life of the loan, but incurs major up-front costs in qualifying the project and the lender and putting the loan on the books. Therefore if the loan is paid off prematurely at an early date the bank loses money. For this reason the commitment provides for a penalty of 7% of the face of the mortgage to be paid to the bank if the mortgage is paid off during the first year. The penalty declines 1% per year if the loan is paid off in any subsequent year. For example, if the loan is paid off during the second year the penalty is 6%, declining each year to zero penalty if the loan is paid off after the seventh year.

6.11.   Escrow

The escrow paragraph of the commitment requires the borrower to make the deposits for taxes and insurance with each installment. This means that every month, the owner, in addition to paying principal and interest on the mortgage, must pay an additional monthly amount sufficient to cover the taxes and the insurance when they come due. If, for example, the real estate taxes on this project are $36,000 per year, then $1/12 \times \$36,000 =$ $3000 additional will have to be paid each month into an escrow account. In addition, if the insurance is $2400 per year, the borrower will have to pay $1/12 \times \$2400 = \$200$ additional each month for insurance. This total of $3200 will be paid to the bank and put into a special escrow account. An escrow account is one that cannot be used by anybody for any purpose except the stated purpose for which it is intended. It cannot be used to cover delinquent mortgage payments. It cannot be used for anything except the taxes and insurance. Some mortgage lenders may also require monthly escrow payments for assessments (such as paving or sewer assessments) and sometimes even for anticipated future expenditures such as replacements for hot water heaters, carpets, and so on.

The escrow account requirement is made as protection for the lender's collateral. Regardless of what mortgages are recorded, real estate taxes always take first priority. Even if all mortgage payments are made on time, if the taxes are unpaid the government can foreclose on the property and have it sold at auction. The proceeds of the sale are used to pay first the back taxes and then the mortgage. For this reason the lender insists that sufficient money be in the bank when the taxes come due. The necessity of fire and extended coverage insurance is obvious, as is the need to have the money on hand to meet these premiums.

6.12. Additional Terms and Conditions

After designating the basic terms of the commitment for mortgage the bank adds seven numbered paragraphs of additional terms and conditions.

6.13. All Costs in Connection with This Loan

Paragraph 1 lists several costs that are incurred in lending the mortgage money. These costs could be paid by the bank and recovered by special charges or higher interest rates. It is simpler just to let the borrower pay them as specified in this paragraph. The bank requires that the owner pay not only his own legal fees (Browns and Associates) but also the fees of the bank's attorney.

6.14. Title Insurance

Title insurance is a single-payment insurance policy. The premium is collected only once and will amount to about a third of 1% of the mortgage.

It must be paid in full before any construction can start. The title insurance company will search the public records for flaws in the title. They will then exclude from the coverage of the policy any listed easements, liens, and so forth that they might find and insure the title against any undiscovered errors in the public record. The title company is expert in examining the abstract of title, which goes along with the real property every time it is sold. They trace the ownership from the present to sometime in the past, usually as far back as several hundred years. Occasionally an error occurs in recording a transfer of title and that error may cause a cloud on the title in the future. For instance, suppose that the records show that in 1925 John Abernathy gave a half interest in the land to his brother, Joe, who promptly moved to Australia. Now, after construction is completed, along comes Joe's grandson, who claims half ownership of the land and won't settle for less than $500,000. The title insurance company may be responsible within limits, although there are usually many caveats in the insurance policy. For instance, the limit of liability is usually specified as the value of the land when purchased. Furthermore the insurance company usually excludes responsibility for claims not on public record, so if the transaction were not recorded the coverage could be excluded. Thus, even with title insurance a good attorney will search the records for any evident flaws.

Exceptions in the title policy might include such things as sale of mineral rights or an easement once granted or an insignificant encroachment such as a fence or hedge encroaching 3 inches onto the property. It may be that the title is acceptable even with these exceptions in which case the bank's attorney could acknowledge the minor flaws but accept them as they are. However, if the exceptions are serious, the bank has the right to refuse to make the loan. These exceptions should show up when the search for title is made by the attorneys, the title company, or by the mortgage insurance company.

6.15. Survey

A survey is another necessary cost paid by the borrower. A survey is especially important when constructing improvements on undeveloped land. For instance, in the case of buildings subject to statutory setback distances an accurate survey of the property line is needed in order to locate all buildings properly. If a building has not been set back far enough from the property line, the building inspector may be legally prohibited from issuing the owner a certificate of occupancy, meaning that no one can move into or use the building.

In addition a survey will reveal any encroachments on the property, whether the actual property is of the same configuration as the description, as well as the accurate location of the property corners.

6.16.   Appraisal

The appraisal is mentioned in both Paragraph 1 and Paragraph 6, which require that the borrower provide an appraisal of the value of the project done by a qualified professional, that is, a Member of the Appraisers Institute or of the Society of Real Estate Appraisers. The bank needs assurances that the project has been built in accordance with plans and specifications, that the quality of workmanship meets the standards of the trade and that the end result has a market value in line with the original estimate.

6.17.   Recording Charges

The mortgage and supporting mortgage note are recorded in the records of the local county courthouse. The County Clerk charges a fee for this which is paid by the borrower. In addition, there are one-time documentary stamp taxes levied on the value of the transaction. These stamps are placed on the mortgage document before it is filed. These fees, together, can amount to several thousand dollars and are paid by the borrower. They must be paid before the mortgage money changes hands.

6.18.   Closing

The closing of the loan is the final step. It is the time when all money changes hands and all legal documents are signed and delivered by all parties. In some cases there may be two closings. First there would be an initial closing when construction is about to begin, and then a second and final closing when construction has been completed. In the commitment document under consideration the Savings & Loan acts as both construction lender and final lender. If the construction lender is different from the permanent lender, the permanent lender generally buys the mortgage from the construction lender. This prevents the duplicate payment of recording fees and documentary stamp fees.

6.19.   Late Fees (Paragraph 2)

The bank is committing to a loan at competitive market interest rates taking into account the reduced risk provided by securing two guarantors of the mortgage. Without these personal signatures and without the reputation and financial stability of the two persons involved, the interest rate necessarily would be somewhat higher. The bank also wants to encourage timely payments since late payments reduce the already thin margin of profit. Therefore, a late charge of 4% is stipulated. This is not an interest rate but simply a penalty fee to be paid in the event the monthly payment is late. For example, if the payment were $1000 per month then the late charge would be $40. To further encourage timely payments the bank stipulates the right to raise the interest rate by 1% per annum "from the time we have a right to accelerate the indebtedness." This "right to accelerate" is usually contained in the mortgage note that accompanies the mortgage and generally occurs 30 days from the due date of a delinquent payment. The "right to accelerate" means the right to declare the mortgage balance due and payable upon default on monthly payments. Thus if a monthly payment should become overdue by 30 days or more, the bank could declare the whole mortgage balance due and charge an extra 1% until it was collected.

6.20.   *Approval of Attorneys (Paragraph 3)*

This paragraph reserves the right for the bank to cancel the commitment if any of a wide assortment of irregularities are detected by the bank's attorneys. In addition, the bank must be assured that the borrower has good title to the property since the property is the primary collateral of the loan. If the title turned out to be seriously flawed the bank's chances of receiving repayment on their mortgage loan would be very poor. Mechanic's liens are claims by contractors, suppliers, laborers, and others who worked on the project that they have not been fully paid. Most state statutes provide at least some circumstances under which a mechanic's lien may take priority over a first mortgage, so the lender bank is understandably concerned that any such liens be cleared up before disbursing any mortgage loan funds.

The last provision of Paragraph 3 is that the borrower provide a final current survey of the completed project. This is needed to insure that the project is located on the property in conformance with the plans, and that there is no violation of local land use, zoning or setback ordinances. As noted previously, even an inadvertent violation of these regulations could render the project untenable and useless.

6.21.   *Insurance (Paragraph 4)*

In this paragraph of the commitment the lender requires that the property being mortgaged will be insured against loss by fire and other hazards as may be reasonably required by the lender (including flood insurance if applicable). The insurance must be in an amount and with a company or companies satisfactory to the lender bank, and the borrower shall furnish extended coverage. Generally, any reputable company will be satisfactory to the lender. Coverage should include liability for personal accidents that may occur after the project is completed. For example, if someone were to slip and fall, and as a result sues the project, then adequate insurance to cover such an accident should be available. With damage awards currently rising to large sums, full protection against such contingencies is a necessity. The "other hazards" envisioned include such subjects as wind damage or earthquake.

6.22. *Certificate of Occupancy Required (Paragraph 5)*

Most local governments have adopted land use and zoning laws for the protection and enhancement of the quality of life in the community. Each parcel of land under the jurisdiction of the local governing body is zoned for certain restricted uses. In some cases, the zoning categories are quite liberal and may consist of broad categories such as residential, commercial, and industrial. However, in many areas zoning has become quite restrictive and as many as 30 or 40 different zoning categories may be encountered within one small city. If a proposed project is known to violate a zoning restriction a building permit will not be issued. However, a permit can be issued in error, either inadvertently or under false pretenses. Or a valid building permit could be issued and the project could be constructed differently from the design shown on the plans. For example, an error in measurement of the setback line could cause the building to seriously encroach upon the setback distance. Inadvertent minor violations can usually be excepted by the Board of Adjustment, but gross violations are seldom tolerated. If a serious violation of the zoning rules occurs, then the owner may not be permitted to occupy the building. In such a case, the building is useless and the bank would not have anything of value as collateral for its money. Therefore, the bank requires that evidence be given of conformance with zoning, land use and deed restriction requirements.

A certificate of occupancy not only acknowledges apparent compliance with zoning and land use requirements but also with all applicable building codes. During construction the building is inspected periodically by inspectors from the Building Department of the jurisdiction. Usually there are several inspectors, each one a specialist, such as the plumbing inspector, electrical inspector, and a structural inspector. These various inspectors make sure that the construction is proceeding in accordance with the plans and specifications for which the permit was issued and in accordance with the governing building code. Recently some cities and counties have been the subject of legal suits alleging that certificates of occupancy were issued and that subsequently the building proved to be faulty. This legal activity will likely result in more stringent requirements for inspection of construction in the future. The bank requires the borrower to obtain a certificate of occupancy in order to satisfy the bank that the construction is legally acceptable, that the zoning and land use regulations have been adhered to, and that there are no known impediments toward occupying the building for its intended productive use.

6.23.   Appraisal (Paragraph 6)

The appraisal is required to be sure that the project's actual market value equals or exceeds the previously estimated value. This was discussed in Section 6.16.

6.24.   Reasonable Application (Paragraph 7)

In Paragraph 7 the bank is attempting to assure the borrower that no unreasonable demands or interpretations will be made regarding the various items in the commitment. Although the bank reserves the right of approval, both for itself and its attorneys, and a satisfactory level of performance is expected, the bank is assuring the borrower that it will not impose unreasonable demands. The bank will close the loan if the borrower performs in accordance with the contract in an honest and straightforward manner.

6.25.   Permanent Loan (Paragraph 8)

Paragraph 8 of the commitment indicates that this commitment shall survive the execution and delivery of the note, mortgage, and all other documents. This simply implies the requirements of the commitment are not superseded by the terms of the note, mortgage, and other documents unless specifically declared to be so. Therefore the borrower cannot ignore the stipulations required by the commitment unless specifically released from them by subsequent documents.

Paragraph 8 also states that this commitment is not assignable. In other words, the commitment is to Maple Hill, Limited and Robert and Harry Metcalf because of their background, record of experience, managerial personnel, and financial responsibility. In this respect a commitment is not a piece of merchandise that can be bought, sold, or traded. It is evidence of an agreement or contract that is to be made between two specific parties, and only those two, unless the agent for the lender approves of a substitution for the borrower. This clause is inserted to eliminate promoters who put together a package and then sell it before construction without any substantial investment or involvement on their part, leaving the lender with a potentially weak developer. The loan is not being made to the site or to the project but to the developer, who has himself been approved by the bank.

6.26. *Commitments from Other Lenders*

The commitment just described is one given by a Savings & Loan Association. Commercial banks and other lenders may have other requirements for a commitment. These are discussed in the following paragraphs.

6.27. *Construction Loan Assignment*

When a commitment for permanent financing is made by a commercial bank, an insurance company or a similar lending institution, then the construction loan usually is distinct and separate from the permanent loan. With reference to construction, a commitment often contains a clause "by which time the final advance must be made and the loan assigned to." This clause refers to the money coming from the construction lender, which is disbursed monthly in accordance with the progress of the work. From the time construction starts, the job may take several months or even a year or more to reach completion. During all this time no money will be coming from the permanent lender. All the money necessary for construction must come either from the developer himself or from a construction lender. Some banks specialize in such loans.

When the work is completed, the construction lender will issue the balance of his loan that is the final advance. Notice that this must take place on or before the expiration date of the commitment. Shortly thereafter, the construction loan will be incorporated into the mortgage loan. The lender of the permanent financing will pay off the construction loan in full and issue a check for the balance to the borrower. (Normally there will be a balance, because the mortgage loan is always greater than the construction loan.) The borrower then has a long-term debt with the mortgage lender, and the construction loan is cancelled.

Often the permanent lender in his commitment will require that the terms of the construction loan document shall be such as to meet all the requirements in the commitment of the permanent mortgage and insist they are to be complied with on the date of the purchase of the loan. This permits the purchase of the loan without any problems developing. One loan simply slides into the other since the terms and conditions of each are the same.

6.28. *Deadline for Acceptance*

Generally, the lender will give the borrower about one month to decide whether or not the proposed terms are acceptable. If the commitment is only for a permanent post-construction loan, the borrower will have to arrange short-term financing for the construction period. If the short-term financing is unobtainable for any reason, the commitment for permanent financing cannot be accepted by the borrower. Most borrowers can use the commitment for the permanent loan as security to obtain the construction financing.

6.29. Good-Faith Deposit

If the commitment is accepted, the borrower is usually required to deposit about 1 point, that is, 1% of the mortgage, with the lender at the time of acceptance. If the borrower then abandons the project or wants to take a loan from a different bank, this deposit is forfeited. In this case, where the mortgage is $1,200,000 a binder of $12,000 would be deposited with the lender as a guarantee that the borrower is serious and to reimburse the bank for time spent on qualifying the borrower if the loan offer is not accepted. Should the project proceed, this binder becomes part of the two or more points and fees that are usually required to close the loan.

6.30. Time of Completion of Construction

Generally a commercial lender will provide an expiration date for the commitment for permanent financing. This means that the entire project must be substantially completed by then. Up to 1½ years may be allowed for the developer to finish construction of the project. If the construction work is not completed by the completion date, then the lender could alternatively (1) review the case and decide against lending anything, (2) raise the rate of interest, (3) decrease the time for payment, or (4) call for a penalty payment. Therefore, before a decision is made to accept a commitment, a developer should be quite certain that the proposed project can be completed within the stipulated time period, to avoid making a costly mistake.

6.31. No Adverse Change

A commitment will often include a sentence such as "except as may be required by this commitment, the loan, the rental income of the rental property, the credit of the borrower and all other features of the transaction shall be as offered in the loan submission without material adverse change." The project could take a year or more to construct, and this commitment will pledge the bank to lend $1,200,000 at that future date. The lender wants to be sure that when the loan money actually changes hands the borrower will be in a good financial condition. Therefore, the lender stipulates that the financial statement of the borrower reflect a stable or improving financial position, since the credit worthiness of the developer at the time of the application was a major consideration in making the loan.

The loan application would also contain a certain rental schedule for the proposed project which is used in appraising an economic value for the project and a consequent loan value. Therefore, the lender would probably require adherence to the original rental schedule as submitted with the application.

6.32.   Completion

The commitment could speak of plans and specifications for the landscaping as well as for the buildings and other improvements. A good landscape design done by a qualified landscape architect can greatly increase the value of any project. Cases are documented in which intensive landscaping saved the project economically, causing a previously barren project to suddenly attract tenants. A landscaping budget reasonably can reach as high as 10% of the cost of the land or 2% of the cost of the project.

6.33   Lender's Right to Inspect

Some commitments contain the sentence "our representative shall have the right to inspect all improvements periodically during and after construction. Periodic inspections during construction shall be made, and such construction shall be approved by our architect." This indicates that the lender has a right to inspect, not only during construction, but also after it. In other words, should the property be damaged or neglected at any time, the lender reserves the right to come to the project and make sure that everything is maintained in first-class condition.

The lender's representative would make these inspections from time to time, not necessarily daily or weekly. Usually they are made weekly or monthly during construction, and yearly thereafter. The inspections made during construction are done to insure that the plans and specifications that were submitted and approved are being adhered to.

7

Creative Plans
for Mortgage Financing

Nathan S. Collier

Key Expressions Used in This Chapter

AMI: (Alternative Mortgage Instruments) Any current type of mortgage used in place of traditional Fixed Rate Mortgages (FRM).

AML: (Adjustable Mortgage Loan) A Variable Rate Mortgage (VRM) with most of the regulatory restrictions removed.

ARM: (Adjustable Rate Mortgage) See VRM below.

BD: (Buydown) A large initial payment by the builder (or other) is used to reduce the monthly mortgage payments for the first few years.

BOY: Beginning of year.

EOY: End of year.

FHLBB: (Federal Home Loan Bank Board) A federal agency with power to regulate federally chartered Savings & Loan Association.

FNMA: (Federal National Mortgage Association) Buys mortgages from original lenders, sells to investors, and thereby creates a market for mortgages.

FRM: (Fixed Rate Mortgage) A mortgage whose interest rate remains constant throughout the life of the mortgage.

FSLIC: (Federal Savings & Loan Insurance Corporation) A federal agency that insures depositors' accounts in S & Ls.

GPAML: (Graduated Payout Adjustable Mortgage Loan) Similar to GPM but the interest rate may be adjusted to follow the market as with an AML.

GPM: (Graduated Payment Mortgage) Payments start low and increase in predetermined steps each year. The interest rate is fixed. Helps borrowers qualify for higher priced homes.

PAL: (Pledged Account Loan) Has graduated payments like a GPM but uses the down payment to compensate for the negative equity.

PLAM: (Price Level Adjusted Mortgage) The remaining principal balance due is recalculated annually to compensate for inflation.

RAM: (Reverse Annuity Mortgage) Permits the elderly to receive monthly payments using their house equity as collateral.

RRM: (Renegotiable Rate Mortgage, or ROM—Roll Over Mortgage) A mortgage whose interest rate can be renegotiated periodically. The periods are preset and are usually 3, 4, or 5 years in duration.

S & L: (Savings & Loan Associations) Banking institutions whose principal function is to attract funds from individual savers by safeguarding and paying interest on these savings, while lending these deposits out on qualified real estate mortgages at reasonable interest rates.

SAM: (Shared Appreciation Mortgage, or SEP—Shared Equity Programs, EPM—Equity Participation Mortgage, CAM—Contingent Appreciation Mortgage) Below market rate interest mortgages with the lender sharing in the net appreciation of the property value.

VRM: (Variable Rate Mortgage, or VIRM—Variable Interest Rate Mortgage, or ARM—Adjustable Rate Mortgage) A mortgage whose interest rates may change at stated time intervals in response to some index number. The magnitude of the rate change may or may not be limited.

WAM: Wrap Around Mortgage

7.1. Good Financing—A Necessity for a Good Project

Financing terms and interest rates can make or break a development project. Some terms and rates are fine tuned to insure the success of a project, while other terms and rates are harsh enough to ruin an otherwise promising project. The crucial differences in terms and rates stem from the ability of the developer to apply up-to-date innovative and creative financing techniques to the project. The old days of the low interest fixed rate long-term mortgage are almost gone. A number of new more complex financing plans have arisen in its place. The number and types of financing plans available now is large and is limited only by the imagination of the lenders and borrowers. Some typical examples of current mortgage financing plans are presented in the following sections.

7.2. Problems with Fixed Rate Mortgages (FRM)

The savings and loan (S & L) industry is still the major source of funds for home loans in the United States and from about 1935 to 1975, relied heavily upon the Fixed Rate Mortgage (FRM) as a vehicle for making these loans. The S & Ls serve as financial intermediaries, borrowing from depositors and lending to builders and home buyers. Traditionally, S & Ls have used their depositors' funds (which could be withdrawn on short notice) to lend to builders and home buyers under long-term loan commitments which could *not* be called in even if the short-term depositors wanted to withdraw their deposits that were funding these long-term loans. In other words, the S & Ls borrowed short and lent long using FRMs. Obviously the FRM was intended for a stable economy, with no worse than mild inflation and predictable interest rates. While interest paid to depositors by S & Ls may fluctuate with the market, under FRMs the interest received by S & Ls from mortgages is at a fixed rate. Thus, in times of rising interest rates, depositors demand an increasingly higher return on their deposits or they will withdraw them and place their funds elsewhere. The FRMs made by the S & Ls previously at lower fixed interest rates do not provide a return sufficient to allow the S & Ls to pay competitive rates to depositors in order to retain old money and attract new money to continue the loan cycle. This causes a process known as *disintermediation* as depositors' funds are withdrawn from S & Ls to flow to higher yield investments. This leaves S & Ls without sufficient funds to finance home building and home buying.

To combat disintermediation, S & Ls attempt to charge higher mortgage interest rates in order to increase the rate of return on their loan portfolios. Higher income enables them to pay higher rates to depositors and thus attract new depositors in order to get new funds to lend. However this can create its own problems. Since by definition the interest rate on a previously made FRM cannot be changed, the only way for an S & L to increase income is to raise the rate charged new borrowers. Thus new borrowers in an inflationary era end up subsidizing old mortgagors who borrowed with an FRM in preinflationary times. Even this inadequate alternative is further constricted if higher lending rates chase away potential mortgage borrowers. S & Ls have experimented with several ways to attract new depositors. At one time Congress authorized S & Ls to issue six-months certificates of deposit with new higher interest rates tied to the U.S. Treasury Bills short-term interest rate. In some respects this only made the situation worse.

Over half of the funds in these new accounts were not new money, but resulted from old depositors transferring money out of existing savings accounts which paid lower rates.

In order to raise cash to make new loans, S & Ls often sell their existing older FRM's in the secondary market. This can also create problems. An old 7% mortgage with a balance of $10,000 will not sell for $10,000 in a secondary market that demands a rate of return of 14%. Instead the $10,000 mortgage has to be discounted and sold for say $7000. The S & L then loses money on the sale which it has to try to make up with even higher interest charges on new mortgages.

Thus, the housing finance market is under pressure from two directions: 1) S & Ls face a cost of funds that is both high and volatile. Therefore, S & Ls need a loan portfolio with a rate of return that is either high enough to cover all foreseeable contingencies 30 years into an uncertain future or flexible enough to move rapidly with changes in the cost of funds. 2) On the other hand, home buyers need creative financing that will be innovative and flexible enough to be tailored to a buyer's specific needs and financial ability. The FRM does *not* take into account the rising income patterns of most young adult home buyers, nor the leveling off of income in the middle years and the decline in income of the retired.

7.3. New Methods of Financing—Alternative Mortgage Instruments (AMI)

In response to these pressures, a host of new financing methods have emerged to meet these needs. Collectively, they are often known as Alternative Mortgage Instruments (AMIs). There are two basic types and then combinations thereof. The first is really just a convenience device designed to restructure the repayment schedule so as to make the initial payments smaller and thus the initial purchase easier for the buyer. The Graduated Payment Mortgage (GPM), Pledged Account Loan (PAL), Wrap Around Mortgage (WAM), and the Buydown (BD) are prime examples of the first type. The second type varies the interest rate or principal amount by an inflation-tied index thus allowing the lender to maintain a constant real rate of return in inflationary times. The Adjustable Mortgage Loan (AML), Price Level Adjusted Mortgage (PLAM), and Shared Appreciation Mortgage (SAM) are good examples of the latter type. Hybrids of the two types include Reverse Annuity Mortgage (RAM) and the Graduated Payment Adjustable Mortgage (GPAM).

7.4. Variable Rate Mortgage (VRM)

The Variable Rate Mortgage (VRM), sometimes known as a Variable Interest Rate Mortgage (VIRM) or the Adjustable Rate Mortgage (ARM),

first came into widespread use in the mid-1970s by state chartered institutions in California. California, with 11 of the largest 15 S & Ls in the
U.S., has long been a leader in real estate financing innovation. California
was among the first of the states to authorize AMIs for its state chartered
S & L institutions. In 1979 the FHLBB authorized all federally chartered
S & Ls to offer VRMs. Since many states have tie-in provisions extending
to state chartered institutions the same powers as federal institutions, this
action had a significant ripple effect.

A VRM is simply what its name implies: a mortgage whose interest rate
varies. The key factors are 1) what index the rate is tied to, 2) how often
the rate can change, and 3) the limits, if any, on how much the rate can
vary both in terms of each incremental change and the total change over
the life of the loan.

In terms of state chartered institutions, the answers to these questions
are as varied as the imaginations of their respective legislatures. Federal
regulations have allowed a maximum of one interest rate change per year
except for the first year when no change is permitted. The minimum rate
change is $1/10$ of 1%, the maximum change at any one time is $1/2$ of 1%. The
maximum rate increase over the loan term is 2.5%. Upward changes are
permissive, downward changes are mandatory. Detailed notices and dis-
claimer provisions are also required. No more than 50% of the dollar
amounts of loans made in any one year could be VRMs. Borrowers must
also be given the option of a FRM, though not necessarily at the same interest
rate.

7.5. Renegotiable Rate Mortgage (RRM)

The Renegotiable Rate Mortgage (RRM), sometimes known as the Roll Over
Mortgage (ROM), is very much like the VRM except that the rate changes
are much more infrequent. In 1980, the FHLBB authorized Federal
S & Ls to offer RRMs, which are loans of up to 30 years but with terms of
3, 4, or 5 years. At the end of each term the loan is automatically renewable
without alteration save as to interest rate. The interest rate may fluctuate
a maximum of 5% up or down over the life of the loan. The maximum
permitted change for each term is 0.5% per year of the term, that is, a note
with 3-year terms can fluctuate 1.5% per term, a note with a 5-year term
may fluctuate a maximum of 2.5% per term.

While the VRM and RRM were improvements on the FRM as far as
allowing an S & L's loan portfolio rate of return to adjust to changes in
the cost of funds, in times of rapid inflation yields on loan portfolios still
lagged significantly behind the short-term cost of funds. At most a VRM
can be adjusted by 0.5% per year and a RRM by 1.5% every 3 years. Yet
the average effective mortgage rate on new homes has jumped as much as
4% in one year.

Example 7.1.   Comparison of Three Types of Mortgages. Find the Monthly Payments Required for a Fixed Rate Mortgage (FRM), a Variable Rate Mortgage (VRM), and a Renegotiable Rate Mortgage (RRM).

Assume a $50,000 loan with a 30-year amortization. Assume the interest rate on the conventional FRM is 12% and the corresponding interest rate on AMIs such as the VRM and RRM is 10.5%. The difference in interest rates reflects the inflation risk premium that is added by the lender to the FRM interest rate. With the VRM and RRM most of the risk of inflation is borne by the borrower, not the lender. Then assume that inflation causes interest rates to rise 0.5% per year over a 10-year period. The FRM rate cannot change. This typical VRM can rise with inflation every year but cannot exceed a 0.5% increase per year, nor can it increase more than a total of 2.5%. This typical RRM can rise with inflation but is adjusted only once every 3 years. The adjustment cannot exceed 0.5% per year (1.5% per 3 years) nor exceed a total increase of 5%. In the later years of this 30-year loan the FRM turns out to be the worst loan for the lender and the best buy for the borrower. However during the first 10 years the total cash flows differ by less than 3% between best and worst while the present values differ by less than 1% over the first 10 years. Note that the opposite result would occur if interest rates fell instead of rose.

Solution: Under the provisions outlined above, the monthly payments are calculated as shown in Table 7.1

7.6.   *Adjustable Mortgage Loan (AML)*

In response to lenders' complaints about a lack of flexibility, the federal government replaced Variable Rate Mortgages (VRM) and Renegotiable Rate Mortgages (RRM) with the Adjustable Mortgage Loan (AML) for all federal S & Ls effective in 1981. The AMLs were truly conceived in the light of the Reagan Administration's push for deregulation, for an AML is merely a VRM with most of the regulatory restrictions removed.

The AML has no restrictions on (1) the frequency of rate change, (2) the maximum amount of any individual rate change, (3) the aggregate maximum rate change, (4) the percentage of a S & L portfolio that can be composed of AMLs, (5) no requirement that the lender offer a FRM alternative, and (6) no restrictions on the use of negative amortization, where the monthly loan payments are insufficient to cover the interest due. The unpaid excess interest is then added to the principal balance owed on the loan.

The only requirement in the choice of an index is that it be readily

TABLE 7.1.

Year	Fixed Rate Mortgage[a] Interest Rate	Fixed Rate Mortgage[a] Monthly Payment Amount	Variable Rate Mortgage[b] Interest Rate	Variable Rate Mortgage[b] Monthly Payment Amount	Renegotiable Rate Mortgage (3-year term)[c] Interest Rate	Renegotiable Rate Mortgage (3-year term)[c] Monthly Payment Amount
1	12%	514.31	10.5	457.37	10.5	457.37
2	12%	514.31	11.0	476.16	10.5	457.37
3	12%	514.31	11.5	495.15	10.5	457.37
4	12%	514.31	12.0	514.31	12.0	514.31
5	12%	514.31	12.5	533.63	12.0	514.31
6	12%	514.31	13.0	553.10	12.0	514.31
7	12%	514.31	13.0	553.10	13.5	572.71
8	12%	514.31	13.0	553.10	13.5	572.71
9	12%	514.31	13.0	553.10	13.5	572.71
10	12%	514.31	13.0	553.10	15.5	652.26

[a]No change in interest rate allowed.
[b]One-half of 1% per year change in interest rate allowed, up to a maximum of 2.5%.
[c]One-half of 1% per year change in interest rate per term allowed, up to a maximum of 5% over the life of the loan.

verifiable and beyond the control of the lender. The FHLBB does suggest several indices:

1. FHLBB's District Cost of Funds to FSLIC insured S & Ls
2. The national average contract mortgage rate for the purchase of existing homes
3. Three-month and six-month Treasury bill auction rate
4. The yield on Treasury securities adjusted to constant maturities of 1, 2, 3, or 5 years.

The FHLBB also specifically preempted under the federal supremacy doctrine all state laws directly or indirectly restricting AMLs issued by federal institutions. Such state laws typically put limitations on variable rate loans or on the charging of interest on interest which commonly happens on a loan with negative amortization. Since being superseded by the AML, VRMs and RRMs as such are no longer offered by federal S & Ls, but many still exist. The AML basically leaves the determination of loan parameters up to the marketplace. Competitive pressures and the need for acceptance by the public should cause meaningful limitations to be put on adjustments. It should be noted that most all the FHLBB restrictions on federally insured

S & Ls apply primarily to mortgage loans secured by one-to four-family dwellings. The conditions and terms of loans on commercial and multifamily units already are determined by the competitive pressures of the market-place. The Comptroller of the Currency, regulator of national banks, has issued regulations that are similar to the FHLBB regulations on AMLs but which retain several restrictions including a limit on the number of rate changes.

7.7. Price Level Adjusted Mortgage (PLAM)

The basic motivations of S & Ls in their move away from FRMs is their inability to reliably predict the effects of inflation on their costs of funds over the life of a 25-to 30-year loan. The typical FRM is amortized over the life of the loan with much cheaper inflated dollars. That is, the pur-chasing power of a dollar a lender lends now is much greater than the purchasing power of the repayment dollar paid back as much as 30 years later. A Price Level Adjusted Mortgage (PLAM) alleviates the S & Ls' con-cerns by neatly shifting all risk of inflation to the borrower. A PLAM has a fixed rate of interest that does not reflect any anticipated inflation. Instead, the outstanding principal amount is adjusted periodically to reflect inflation. For example, borrow $10,000 now. If inflation is 10% over this coming year, then at the end of the year the principal balance remaining is adjusted upward 10%. The new principal balance to be paid is 1.10 times the old principal balance. At the same time the monthly payment also is adjusted upward in order to amortize the additional principal over the remaining life of the loan. The advantages of PLAM are many. Since the lender no longer has to add inflation to the interest rate, the interest rate charged tends to be low. The required real rate of return for the use of dollars, exclusive of inflation, might be as low as 4.5%. A PLAM reflects only actual inflation, not worst case anticipation of inflation. A PLAM will work in times of both inflation and deflation. In times of little or no inflation, it will behave like a FRM. The PLAM can be tied to a variety of inflation related indices. For instance, if the borrower is a wage earner it can be tied to the wage index. Therefore, the borrower's income could be expected to rise at the same rate as the PLAM payments and payment increases should not prove an undue burden. While the PLAM has not been widely au-thorized in the United States, it has enjoyed widespread usage in Brazil since the mid-1960s. Its popularity there is attributed to Brazil's experience with inflation rates of over 100% per year and the consequential necessity for tying promises of future payments in Brazil to various inflation indices. The chief difficulty in introducing PLAMs into widespread usage in the U.S., apart from potential legal problems to be discussed later, may be one of public acceptance of what is a fairly complex and abstract concept. With the recent introductions by the FHLBB of the extremely flexible AML, which will achieve many of the same end results for lenders as a PLAM, much of the need for PLAMs has dissipated.

7.8. Shared Appreciation Mortgages (SAM)

In 1980, the FHLBB approved the concept of Shared Appreciation Mortgages (SAMs). SAMs have also been called Shared Equity Programs (SEPs), Equity Participation Mortgages (EPMs), or Contingent Appreciation Mortgages (CAMs). SAMs are mortgages at below market interest rates but with the lender sharing in the net appreciation of the property at the earlier of (1) maturity of the mortgage, (2) transfer of the property, or (3) full repayment of the loan. The lender's share of the net appreciation is known as the lender's *contingent interest*. The FHLBB regulations limit the lender to a maximum of 40% of the market appreciation and provide for a maximum term of 10 years (but amortized at a rate of up to 40 years) for the SAM. Furthermore, the lender must agree to refinance the SAM at the end of its term. The guaranteed refinancing need not be a FRM, merely anything other than a new SAM. No new fees may be charged for the refinancing. Presumably, the lender must refinance even if the borrower has fallen below the current loan standards.

Net appreciation is defined as any excess after subtracting from the market value of the property (1) the cost of the property to the borrower, plus (2) the cost of capital improvements made. The market value may be set by either the net sales price in case of a sale or transfer, or by an appraisal by a qualified appraiser.

Although the homeowner may gain in the short run with lower fixed interest rates, he may lose in the long run in that the lender's share of appreciation may amount to substantially more than conventional interest charges.

SAMs have several foreseeable pitfalls for lenders also. First, lenders might be eager to offer SAMs in neighborhoods with good appreciation potential, but federal anti-redlining provisions would restrict lenders from offering SAMs *only* in those areas and would require them to be offered also in districts where little or no appreciation could be expected. Second, lenders have a great need for current income. While SAMs offer the potential for an above average rate of return in the long run, in the short run the cash inflow is below that of other AMIs. Thirdly, the most crucial variable for a lender in evaluating SAMs is forecasting the estimated amount of appreciation of the property under the life of the loan. This is primarily dependent upon a forecast of inflation and it is the inability of S & Ls (or anyone else for that matter) to do this accurately that started the move to AMIs.

Example 7.2.   Shared Appreciation Mortgage (SAM). Find the Effective Interest Rate.

Assume a $50,000 property and 5% per year increase in property value, compounded annually for 10 years.

Solution: This yields a property value of $F = \$50,000 \times 1.05^{10} = \$81,445$ at the end of Year 10 for a gain of $81,445 − $50,000 = $31,445. This gain represents the net appreciation if no capital improvements were made. If the lending S & L took the maximum 40% share of net appreciation allowed by FHLBB regulations, the S & L would receive 0.4 × $31,445 = $12,578, that is, 40% of the $31,445 gain. If the initial interest rate is 10% on the $50,000 mortgage for 10 years, then the added payment of $12,578 at EOY 10 raises the effective interest rate to 12.2%. If the initial mortgage interest is higher, then the change in the interest rate caused by the lump sum at EOY 10 is less. For instance an initial rate of 13% jumps only to 14.8%. Obviously the initial interest rate on a SAM would have to be significantly below other rates in order to make this type of loan attractive to a property owner. If 5% appreciation were anticipated then about a 2% interest rate gap on a 10-year mortgage would be suitable for the borrower. Higher anticipated appreciation rates would warrant larger discounts.

7.9.   Graduated Payment Mortgage (GPM)

Unlike many other AMIs, the interest rate of a Graduated Payment Mortgage (GPM) is fixed throughout the life of the loan. This reflects the GPMs basic purpose which is to help upwardly mobile borrowers qualify for a home loan that they would not otherwise be able to afford. When considering a loan, most lending institutions require that not more than 30 to 35% of the income of a family be spent on housing. The GPM helps borrowers qualify by starting out with low payments which then increase in fixed gradients of $x\%$ per year for y number of years. The two important varibles unique to GPMs are 1) the amount of increase in the payment each year and, 2) the number of years over which payments increase. GPMs do nothing to shift the risk of inflation-induced interest rate fluctuations from the lender to the borrower. Therefore, they do not act to relieve the liquidity or disintermediation pressures on lenders. GPMs are primarily a marketing device to qualify people with good future income prospects for mortgage loans on higher priced homes via a convenient restructuring of the payment schedule.

Example 7.3.   The Graduated Payment Mortgage (GPM). Find the Annual Amortization Schedule for a GPM and Compare with an Equivalent FRM.

TABLE 7.2.

| | Fixed Rate Mortgage | | Graduated Payment Mortgage | | | |
Year	Monthly Payment Amount	EOY Mortgage Balance	Monthly Payment Amount	Monthly Interest Due	Yearly Amortization	EOY Mortgage Balance
1	$514.31	$49,819	$412	$500	($1,056)	$51,056
2	514.31	49,614	443	511	(816)	51,872
3	514.31	49,384	476	519	(516)	52,388
4	514.31	49,124	512	524	(144)	52,532
5	514.31	48,832	550	525	300	52,232
.	.	.	.	.		.
.	.	.	.	.		.
30	514.31	0	550			0

Assume a $50,000 loan at 12% amortized over 30 years, with 7.5% per year increases in the mortgage payment over the initial 5 years and then a stable payment amount.

Solution: The comparison between the two amortization schedules is shown in Table 7.2.

7.10. Negative Amortization

In order to reduce at least the beginning payments on a' self-amortizing mortgage loan, it is necessary to resort to negative amortization. This is because the initial payments on long-term loans are almost all interest. Extending the pay-back period beyond 30 years is not an effective way to reduce initial payments because the effect on the monthly payment amount is so small. Negative amortization occurs when the payments on a loan are insufficient to cover the interest due. The unpaid excess interest is added to the principal balance already owed. Thus the principal balance actually rises in the first few years of the loan before the effects of the increasing payments every year begin to reduce the principal balance. The first significant use of GPMs was in 1978, when the Federal Housing Authority (FHA) authorized use of the five different versions of the GPM described below.

Number of Years during Which Payments Increase	Percentage by Which Payments Increase Each Year
5	2.5%
5	5 %
5	7.5%
10	2 %
10	3 %

Plan 3 has proven to be most popular because it provides for the steepest increases and therefore the lowest initial payments. The loan balance is not permitted to rise to more than 97% of the original appraised value of the property. Since FHA's original loan-to-value ratios typically run as high as 95%, in order for the GPM to operate effectively the down payment should be higher than under the normal FHA loan. However, this requirement for a higher down payment tends to defeat the purpose of the GPM program. Buyers who can afford larger down payments usually are not those most in need of the loan-qualifying assistance provided by the GPMs lower initial payments. Fears that GPMs will not serve their intended beneficiaries are also prompted by the lack of upper income limits on those who can apply for the program. In response to these concerns, FHA issued regulations for two additional, more liberal, GPM plans.

Number of Years during Which Payments Increase	Percentage by Which Payments Increase Each Year
5	7.5%
10	4.9%

More significantly, the unpaid balance of the loan may increase to the lesser of 1) 113% of the original appraised value of the property or 2) 97% of the appraised value after increases of 2.5% per year. The new program contains two major eligibility restrictions: Only borrowers who 1) have not owned a home in the past 3 years and 2) who cannot qualify for any other FHA program, can qualify for the new program. The FHLBB has authorized federal S & Ls to offer GPMs with negative amortization. The FHLBB then liberalized its regulations for federal S & Ls by removing the restrictions on the amount of the percentage increases and allowing them to increase as often as monthly. The graduation period of the loan is still limited to a maximum of 10 years. The FHLBB liberalization does not apply to the FHA's GPM program

Under the FHLBB's GPM program the borrower, if he is eligible at that time, may elect to switch to a FRM at the then current rate without any additional charge. The loan-to-value ratio is limited to 95%. The lender must give the borrower a side-by-side comparison of the GPM to FRM. The borrower must be told of the availability of FRMs.

The FHA's GPM program has proven popular. Within a year of its inception about 30% of the single-family loan applications received by FHA nationwide and over 50% from California were for GPMs. While the primary consideration of borrowers in shopping for financing seems to be the amount of the initial payment, in the long run a GPM costs the borrower more in interest charges than a comparable FRM. This is because the negative amortization features of GPMs result in a greater average outstanding principal balance over the life of the loan than under a comparable FRM. Simply put, the borrowers pay for the convenience of the restructuring of the payment stream. Interestingly, GPM borrowers use the program to purchase more costly homes. The implied implication behind the GPM is that housing prices will continue to rise. If housing prices decline, the value of the property may well be less than the amount of the loan. This would probably have an unfavorable impact on the default rate. In the case of a sale, the borrower may be compelled to pay the difference between the depressed sales price and the loan amount. A relatively minor difference between the loan and sales price can quickly become significant. For instance, just the services of a realtor in marketing the property can cost 6% of the sale price. The GPM also inherently assumes that the income level of the borrower will rise sufficiently to cover the increase in loan payments. However, the GPM payments will increase regardless of the trend of the borrower's income or the economy in general. This is one area in which a PLAM tied to a wage index may turn out to be superior.

7.11.   Pledged Account Loan (PAL)

The Pledged Account Loan (PAL) has graduated payments like a GPM but eliminates the negative amortization problem of a GPM. This is accomplished by depositing the money normally used as a down payment into an interest-bearing account with the lender instead of using the down payment to create equity in the property. Then monthly withdrawals of interest and principal are made by the lender from that account and applied to reduce the amount of the monthly payment due from the borrower. The amount in the account is calculated so that it takes about 5 years for withdrawals to exhaust the account. At the end of 5 years the portion paid by the borrower rises to the full amount and the loan is in all respects similar to an FRM. The account is collateral for the loan and is therefore said to be pledged to the lender. Since the down payment is not used to create equity in the property, the amount of the loan will approach 100%. California has approved PALs for loan-to-value ratios of up to 95%.

Example 7.4.   The Pledged Account Loan (PAL).
This example looks at the problem of how to purchase a $100,000 property for $20,000 down with interest at 12% and still keep payments under $520 per month for the first year.

Solution: Use a 95% PAL at 12% interest. The first $5000 of the $20,000 down payment is used as a normal down payment to create equity in the property. The remaining $15,000 is placed in a pledged Certificate of Deposit or savings account earning 8% interest at the lender bank. Monthly payments are withdrawn from the pledged account to make up the deficit in amortization payments on the mortgage. This plan results in the following payment schedule for a PAL Mortgage for $95,000 at 12% for 25 years:

Year	Balance in the Pledged Account at EOY	Monthly Amount Withdrawn from Pledged Account	Monthly Amount Paid Directly by Borrower	Total Monthly Payment on PAL
0	$15,000.00			
1	10,251.72	$481.39	$ 519.17	$1,000.56
2	6,307.99	385.11	615.45	1,000.56
3	3,235.58	288.83	711.73	1,000.56
4	1,106.81	192.56	808.00	1,000.56
5	0.00	96.28	904.28	1,000.56
6–25	0.00	0.00	1,000.56	1,000.56

If the $20,000 had been used as a conventional down payment instead of a PAL, the payments on a FRM of $80,000 at 12% and 25 years would be $842.58 per month. The PAL provides a savings during the first year of $323.41 per month, but the buyer pays $157.98 per month *more* over the last 20 years of the mortgage. The comparison would be more favorable if the pledged account earned an interest rate more nearly equal to the mortgage interest rate.

7.12. Graduated Payment Adjustable Mortgage Loan (GPAML)

One criticism of the GPM is that it did nothing to alleviate the problems of S & Ls in matching the yields of their long-term mortgage loan portfolio with fluctuations in the short-term cost of funds. Ever innovative, in 1981 the FHLBB created the Graduated Payment Adjustable Mortgage Loan (GPAML). The GPAML is similar to the FHBBL's liberalized GPM except that the interest rate may vary with the marketplace in a manner similar to AMLs. There are no limits on the amount by which the monthly payment or the interest rate may vary during the life of the loan.

7.13. Reverse Annuity Mortgages (RAM)

Reverse Annuity Mortgages (RAMs) address the housing problems of American elderly in these inflationary times. Most elderly people live on relatively fixed incomes and, in times of ever-rising prices, they may be forced to liquidate assets in order to maintain their living standards. The largest asset of most elderly is their home. The net house equity of those 65 and over is estimated to be around 100 billion dollars. However, homes are not easy to liquidate. Furthermore, the elderly borrowers may need their homes and may not want to realize their equity by selling. Reasons for not selling include an emotional attachment to the existing home and neighborhood, adverse tax consequences, and potential difficulty in finding suitable new housing. Refinancing often will not work because the elderly borrower's income levels are insufficient to qualify for a conventional mortgage. Also a conventional mortgage requires the borrower to make principal payments that in effect have the borrower buying back the very equity they are trying to liquidate. The goal of a RAM is to allow the elderly to utilize the existing equity of their homes to supplement their income yet still remain secure in their homes.

The two common types of RAMs are the Rising Debt Mortgages and the Fixed Debt with Life Annuity Mortgage. Under a Rising Debt RAM the proceeds of the loan are disbursed to the borrower in equal monthly payments. Interest on the funds disbursed is accrued monthly and added to the outstanding balance of the loan. When the loan reaches either a fixed dollar amount or a predetermined loan-to-value ratio the loan has matured. At maturity the loan, depending on the terms of the mortgage, is either due and payable in full or converts into an FRM which must then be paid off. With a rising debt RAM the "mortality risk" is borne by the borrower. Mortality risk is simply the problem of the elderly borrower who outlives the loan. Even though it is not probable that a 65-year-old borrower will outlive a 15-to 20-year RAM, the possibility will rightfully give potential RAM borrowers second thoughts. Very few will have the funds to pay the loan in full. Conversion of the RAM to an FRM does alleviate the problem slightly, but most elderly borrowers who are in need of a RAM to supplement their income obviously will not have the means to make monthly payments on an FRM. As a result, the eventual effect of the maturing of a RAM is to force the elderly borrower to sell and use the proceeds to satisfy the RAM.

Where the maturity date of a RAM is determined by the loan amount reaching a predetermined loan-to-value ratio, it is possible to extend the life of the loan if a reappraisal of the property shows sufficient appreciation since the origination of the loan. At best, however, this merely postpones the day of reckoning. Another potential drawback of RAMs is that most are structured in terms of a fixed monthly payment. If the ravages of inflation continue, 10 years later that fixed monthly annuity that originally was so helpful may no longer be an adequate income supplement.

The Fixed Debt with Life Annuity RAM shifts the mortality risk from the borrower to either the lender or an insurance company. The Fixed

Debt Life with Annuity RAM is similar to an FRM except that the entire proceeds of the loan are used to purchase an income-producing annuity on the life of the borrower. Most likely an insurance company will be used since they can spread the risk of statistical fluctuations over a larger risk pool than could the lender. A portion of the monthly income from the annuity is used to make the payments on the FRM with the remainder going to the borrower. Thus the borrower is assured of having the loan paid off. The shortcoming is that since rates of returns on annuities are traditionally less than that of mortgages, the income supplement for the elderly borrower will be less than that of a Rising Debt Mortgage. The FHLBB authorized RAMs for federal S & Ls in 1978 but gave no regulation for their formulation, preferring to receive S & L proposals on a case-by-case basis. If the FHLBB fails to object within 60 days, the S & L may proceed with its plan as submitted. The FHLBB did make the following requirements:

A 7-day cooling-off period during which the borrower may change his/ her mind and cancel the application for the loan.

Disclosure of all contingencies that could force sale of the property.

If the loan is for a fixed term, the lender must offer refinancing at the end of the term at then current rates.

Prepayment without penalty.

The loan must have a fixed interest rate.

Detailed disclosure requirements.

In addition, several states have specifically permitted RAMs. Since RAM payments are not 'earned income' they should not affect an elderly borrower's right to receive Social Security or Medicare benefits. Still, an elderly RAM borrower should assess the effect the income from a RAM would have on his eligibility for various other types of financial assistance.

Example 7.5.   Reverse Annuity Mortgage (RAM)—The Rising Debt Version (See Table 7.3.)

Assume a $50,000 initial property value, a 12% interest rate, and 5% per year inflation with a corresponding rise in property values. Further, assume the borrower elects to receive the loan proceeds in monthly payments of $300, that is $3600 per year with the accrued interest and disbursed principal being added to the outstanding loan balance after each payment. Note that in 8 ½ years the outstanding loan balance exceeds the initial property

TABLE 7.3.

Year	Mortgage Balance BOY	Payments to Borrower	Interest Added on to Mortgage Balance	Mortgage Balance EOY	Property Value at EOY with 5% Appreciation	Appreciated Property Value Less Mortgage Balance = Owner's Net Equity Remaining	Buying Power of $3600 with 5% Inflation
1	0	$3.600	$ 216	$ 3,816	$52,500	$48,684	$3,420
2	$ 3,816	3,600	674	8,090	55,125	47,035	3,249
3	8,090	3,600	1,187	12,877	57,881	45,004	3,087
4	12,877	3,600	1,761	18,238	60,775	42,537	2,932
5	18,238	3,600	2,405	24,243	63,814	39,571	2,786
6	24,243	3,600	3,125	30,968	67,005	36,037	2,646
7	30,968	3,600	3,932	38,500	70,355	31,855	2,514
8	38,500	3,600	4,836	46,936	73,873	26,937	2,388
9	46,936	3,600	5,848	52,784	77,566	24,782	2,269
10	52,784	3,600	6,550	62,934	81,445	18,511	2,156
11	62,934	3,600	7,768	74,302	85,517	11,215	2,048
12	74,302	3,600	9,132	87,034	89,793	2,759	1,945

value from year 1 and at the end of year 12 is rapidly approaching the inflation adjusted value of the property.

7.14. The Buy Down

The buy down is basically a marketing device whereby a large payment is made to the lender by someone other than the borrower, typically the builder, the seller or a relative. The payment lowers, or buys down, the effective rate of interest for a specific period of time thus lowering the payments for that same time period. In 1981 the FNMA agreed to begin purchasing mortgages with buy-down provisions with certain restrictions as to how much the interest rate may be bought down and for what period of time. The existence of a secondary market that provides liquidity is a prime prerequisite for the success of any mortgage instrument. It is important to note that these restrictions are placed by FNMA on mortgages it will purchase, not by the FHLBB on mortgage loans federal S & Ls can make. Federal S & Ls would have the authority to make buy downs because the buy downs are merely specialized AMLs and thus the terms would fall within the very liberal FHLBB authorization for AMLs. Note that the same payment which could buy down the interest rate could have been used in a more conventional manner to simply make a larger downpayment. This would act to reduce the amount of the payments over the life of the entire loan and consequently the impact on the amount of the payment reduction is not as dramatic as a buy down which concentrates its effect on the first few years of the loan.

Example 7.6.   Show How a 12% Interest Rate Is Reduced to a 9% Interest Rate for the First 3 Years of the Mortgage.

Solution: Prepay part of the interest by use of a buy down. For example, on a $50,000 loan with interest at 12%, the interest rate can be reduced to 9% for 3 years by prepayment of the 3% difference between 12% interest and 9% interest. Three percent of $50,000 is $1500 per year, which amounts to $4500 over 3 years. However, part of the monthly mortgage payment goes toward reducing the $50,000 principal, causing the monthly interest payment to likewise decline. In addition, the 3% interest buy-down payment is paid in a lump sum in advance and is entitled to a time-value discount. Thus the cost to buy-down the interest from 12% to 9% for the first three years of a $50,000, 25 year loan is only about $3,600 instead of $4,500. The effect of the interest buy-down is to reduce the monthly payments by over $100 per month for the first three years.

7.15.   Wrap Around Mortgage (WAM)

When financing the purchase of an existing property a WAM is one method of incorporating an existing low-interest mortgage into the financing package. When the buyer is not able to afford a large enough down payment to pay for the seller's equity (cash to mortgage), the seller may elect to receive one monthly mortgage payment from the buyer on a WAM. The seller then continues to make payments of a lesser amount to the holders of the existing low-interest mortgages. The balance of the payments is retained by the seller as payment on the seller's equity. Thus a WAM is a mortgage under which a lender has or can obtain funds from one or more low-interest sources and lends them at a higher interest to the WAM borrower. The several sources usually include funds borrowed under one or more existing low-interest mortgages combined with new money provided by the WAM lender. Thus the lender borrows at low interest and lends those same funds together with some new money at a higher interest rate. The WAM secures the sum of the pre-existing mortgage(s) plus the amount of any new financing. The WAM borrower agrees to pay the WAM lender payments sufficient to amortize the sum of the old and new mortgages. In return the WAM mortgage lender agrees to make the payments on all of the pre-existing mortgages. Thus the WAM lender is said to have "wrapped around" the new financing. The attractiveness of a WAM stems from (1) the unattractiveness of a complete refinancing at today's high interest rates and (2) the fact that the amortization period of the WAM is generally longer and the interest rate, while higher than that of the pre-existing underlying note, is still generally less than the prevailing market interest rate. Lengthening the amortization period serves the borrower by reducing the amount of the monthly payments.

Even though the interest on the WAM is below prevailing market interest rates, a WAM is always at a higher interest rate than the underlying mortgage(s). This gives the WAM borrower an advantageous interest rate and also gives the lender an effective yield that is higher than the stated interest rate of the WAM. This is because while the lender receives interest at a higher rate on the full face amount of the note (the old financing and the new financing), the lender has to provide only the new financing. In effect, the lender is borrowing at the old, low interest rate and lending that borrowed money at a higher rate to the WAM borrower. Thus, WAM mortgages are usually only seen in times of rising interest rates. The WAM mortgage should contain provisions specifying the ramifications if either the WAM borrower or the lender fails to make payments as agreed. Typically, the WAM lender's obligation to make payments on the underlying mortgages is conditional upon timely receipt of payments from the WAM borrower. Conversely, a well-drafted agreement will provide that if the WAM lender fails to make the payments on the underlying mortgage(s), the WAM borrower may make the payments directly and receive credit on the WAM. Since the lender's attractive rate of return is dependent upon the interest rate spread, the lender may wish to restrict the borrower's right to prepay down to and assume the pre-existing underlying mortgage.

The WAM mortgage appears to be a situation in which both parties come out ahead. This is possible because it is done at the expense of the lender of the funds that comprised the pre-existing, underlying mortgage(s). Needless to say, it is the reluctance of borrowers to pay off long-term low-interest mortgages, which are commonly involved in WAM mortgage situations, that have contributed to S & L's earnings decline and liquidity problems.

Example 7.7. The Wrap Around Mortgage (WAM). Find Two Ways of Selling a Property and Keeping an Existing Low Interest Mortgages in Effect. Compare with an FRM.

Assume a property with a value of $50,000 and an existing mortgage of $20,000 at 8% with payments of $250 per month. Assume the current market interest rate for FRMs is 12%. A buyer has $10,000 to use as a down payment but wishes to finance the purchase so as not to lose the benefit of the existing low interest rate mortgage.

Solution: The two basic choices are either a second mortgage or a WAM. Both, along with an FRM are illustrated below.

Second Mortgage—Option A: Purchase the $50,000 property with a $10,000 down payment plus a $20,000 second mortgage plus assume the existing $20,000 first mortgage.

$10,000	Down payment
20,000	Assumption of existing first mortgage at 8%; payments $250 per month for 9½ years
20,000	Second mortgage at 12%; payments $220 per month for 20 years
$50,000	Purchase price

WAM—Option B: Purchase the $50,000 property with $10,000 down payment plus a $40,000 WAM.

$10,000   Down payment
$40,000* WAM at 11%; payments of $413 per month for 20 years
——————— No additional closing costs if seller financed
$50,000   Purchase price

Refinance with FRM—Option C: Purchase the $50,000 property with $10,000 down payment plus a new first mortgage for $40,000.

$10,000   Down payment
 40,000   New FRM at 12%; payments $440.43 per month for 20 years
——————— Plus closing costs, typically 5% of loan amount
$50,000   Purchase price

The main advantage of the WAM is that the buyer has lower payments, $413, and a lower interest rate than with a conventional mortgage. The seller can finance the sale at a below market interest rate because while the seller is receiving interest on $40,000 at 11%, the seller has in reality only "lent" $20,000 to the buyer. The remaining $20,000 loaned to the buyer at 11% is borrowed from the existing mortgage lender at 8%. Thus the seller collects the 11% on the $20,000 he actually lent the buyer *plus* the 3% interest rate spread on the remaining $20,000 based upon the difference between the 11% WAM and the 8% pre-existing mortgage. The second-mortgage financing route has the advantage of retaining the benefit of the 8% mortgage but at the disadvantage of higher monthly payments. This is because of its fast pay out, that is, only 9½ years remain on the life of the loan.

7.16. Blended Rates

In an attempt to combat the S & Ls problem of long-term low-interest loans versus short-term depositors demanding higher interest rates, FNMA in 1981 began offering its Refinance/Resale Finance Program. Under that program, FNMA will agree to refinance up to 95% of the property's market value on any low interest rate mortgage it now owns at a blended rate somewhere between the current high market rate and the lower rate of the current mortgage. Interest rate mortgages of below 10% comprise over 80% of FNMA's current loan portfolio. This has resulted in a net mortgage yield of 9.24% at a time when their average borrowing costs are 10.11%. FNMA's blend note refinance program is structured so as to be competitive with

——————

*Remember, with a WAM the underlying mortgage is *not* paid off. Instead, in this example, the buyer would pay the seller payments on the full $40,000 and the seller would in turn make payments on the existing $20,000 mortgage.

WAM rates. Hopefully this will reduce the number of FNMA low-interest loans that are wrapped around and thus remain in FNMA's loan portfolio for longer than originally forecast.

7.17. Usury

Federal law exempts most S & Ls from state-imposed usury limits, however usury limits still apply with full force to transactions between individuals. With a variable rate or shared appreciation mortgage the value of the income to the lender is not known until the mortgage is paid off, thus the lender's income could exceed usury limits unwittingly. Therefore it is advisable to consult a qualified attorney when drawing up mortgage agreements.

8

How to Finance
the Land Purchase

8.1. Finding a Lender

Having secured the option to buy the land, the next step in land acquisition is to find the money to pay for the land. This money is usually borrowed, so the developer must first find a lender who is willing to accept real estate as collateral for a loan or provide the financing for the land in some other way. One conventional source of funds for large-scale land financing is a Real Estate Investment Trust (REIT), a group of investors interested in real estate investments including financing of promising development projects. The REIT may consider a subordinated lease agreement involving purchase of the land by the REIT investors who in turn lease it back to the developer on a long-term (e.g., 40 years) lease. The lease may have to be subordinated to the first-mortgage lenders to provide the security required for a long-term, low-interest, first-mortgage loan.

Other alternatives for land financing include application to banks, mortgage bankers, mortgage brokers, or the FHA for financing for a land purchase. Some banks will write a short-term balloon loan on the land in anticipation of rolling that loan over into a long-term mortgage loan upon completion of the development project.

8.2. Appraisal to Determine Market Value

In order to determine a fair market value on which to base the loan, the land must be appraised. The appraisal can be made by a staff member of the lender's organization or by an independent professional Member of the Appraisers Institute (MAI). The appraisal may result in a different dollar value than the purchase price. In fact, in the case of the FHA, the valuation of the land is based on the market value that will exist *after* development is completed, which predictably is much greater than the value of raw land.

8.3. Loan Application Requirements

An application for a land acquisition loan requires several items. First is the appraisal discussed in the preceding paragraph. Second, a survey must be made by a registered surveyor. Third, it is necessary to present evidence that the zoning is proper for the proposed project. This can be done by presenting either a current zoning map or a letter from the building inspector. Fourth, if utilities are not already in place up to the property line, a letter of intent is needed from the appropriate utility supplier stating that the necessary utilities will be brought to the property by the time the development is completed. In all cases, the lender will require evidence of good title. Since at this stage the buyer has only an option on the land, the title will not be in the buyer's name, but the abstract must show that there will be no trouble in getting title insurance. Finally, where appropriate, it may be necessary to furnish soil borings, designs for retaining walls or seawalls, or anything else that is required by the special conditions of the particular site under consideration.

8.4. Good-Faith Deposit

At the time of application for a land acquisition loan, a bank or an REIT usually will ask for a refundable good-faith deposit, which is the same in concept as the similar deposit made with a mortgage loan application. If the loan is approved but the borrower for some reason refuses to proceed with the project, the lender keeps the deposit to pay for the time and effort spent in processing the application. One purpose of a good-faith deposit is to keep prospective borrowers from shopping around and wasting everyone's time. It stands as evidence of the applicant's sincerity.

8.5. The Loan Memorandum

The loan officer of a bank or REIT will issue a loan memorandum after receiving the loan application and all necessary supporting documents, together with the good-faith deposit. The loan memorandum can actually be issued before the documents are in hand, in which case it is made subject to a satisfactory survey, appraisal, and so forth. In either case, it is simply a descriptive story that spells out, in some detail, what the loan is all about and who the borrower is. The memorandum is given to a loan committee that will either approve, disapprove, or ask for additional information. If approval is forthcoming, a commitment will be issued, which states what additional documents are required. The attorneys of the lender will then write a letter to the borrower stating what additional documents must be submitted. After the borrower complies with all the requirements, the closing of the loan can take place, and the loan money passes from the lender through the borrower (who may not ever actually see the money) and ultimately to the seller of the land.

8.6. *Loan-to-Value Ratio and Interest Rate*

The amount of a land acquisition loan made by an REIT or bank will typically be at least 60% of the appraised value of the site, and can go as high as 75 or 80%. The credit rating of the borrower, together with the type of securities pledged against the loan and the policies of the lender, will determine the amount as well as the rate of interest. A borrower who pledges liquid assets (cash or equivalent) in an amount equal to the face value of the loan in addition to the land, will get more favorable terms than a borrower who offers just the land itself as security. The true rate of interest that is charged may be as much as 4½% over the prime rate. (The prime rate is that which is charged by the biggest banks to their best customers on short-term loans. For example, if Chase Manhattan Bank charges a large and prosperous corporation at the rate of 10% per year on a 90-day note, the interest charged by an REIT on a land loan may be as much as 14½% per year.) The true rate of interest is the effective yield to the lender and includes both interest and discount points because the discount points are equivalent to nonrefundable prepaid interest. The maximum true rate of interest that can be charged including points is determined by state usury laws, as previously explained.

8.7. *Short-Term Loan*

A land acquisition loan is in reality a short-term first mortgage, and as such carries with it a note and a mortgage. The mortgage is secured by the land, and the note is secured by all the available assets of the borrower. Generally the time set for repayment of the principal of the loan is less than 24 months, with an option to renew, although it is sometimes possible to get a land loan for as long as 5 years. The short time limit serves the purpose of insuring that future construction will not be delayed inordinately. A land loan is made in anticipation of legitimate development, and not to enable some speculator to hold a tract while seeking a buyer, hoping to resell it at a speculative profit at some early date. Speculative ventures are far too risky to be financed by a typically conservative bank. Banks much prefer well-researched and secured investments in sound real estate construction ventures and consider them to be reasonable vehicles for the lending of bank funds. Therefore the bank requires that the land be put to use as soon as practical, at which time the short-term bank loans will be repaid by the proceeds of the permanent mortgage when the project is completed.

8.8.   Land Development Financing

After the raw land is purchased for development, site improvements are commenced in preparation for construction of buildings. These on-site improvements include not only streets, sidewalks, sewers, water, and electric lines but also site preparation. Site preparation consists of reshaping the land by changing existing grades through cutting and filling, and removing any trees that stand within the building or pavement lines. All this work is done in accordance with plans made by an architect working with a landscape architect and an engineer. Short-term loans are needed to finance this work also.

8.9.   Land Development Loan Draws

Site preparation on a large project may take several months. Since construction on the building has not yet been started, no draws are available on the construction loan for the building. To finance the site preparation work during this period a land development loan is needed. The purpose of a land development loan is to provide interim financing and to enable a developer to prepare the subdivision site for the start of house building. The money for a land development loan will be paid out in draws, similar to the draws on a building construction loan, each of which has been certified by an engineer at a specified date in each month. The engineer verifies that the work has been performed satisfactorily and in accordance with the plans, and that the value of work claimed in the draw request is proper and correct. After the on-site improvements have been completed, the borrower will be obliged to repay the land development loan in monthly installments, in a manner similar to that for a first-mortgage loan on a building. The interest rate is about the same as that for a land acquisition loan.

8.10.   FHA Title X

To encourage provision of more and better shelter for our people across the nation, the Federal Government, through the FHA, has established several programs that provide loan funds for development projects. One popular and successful program is the FHA Title X program.

The minimum size tract eligible for a Title X loan under FHA financing is 10 acres, and the maximum is 1000 acres. The prinicipal amount of the loan amounts to 50% of the raw land value, plus 90% of the cost of developing the land, up to a maximum combined total of $20,000,000. In addition to the outlay for streets, curbs, gutters, walks, and sewer and water lines, the cost of developing could include such amenities as a recreation center, golf course, tennis courts, and public buildings. Furthermore, the total replacement cost includes professional fees, utility costs, and interest on the loan during the repayment period, and all of these items are also included in the development cost when computing the principal amount to be loaned.

8.11.   Title X Terms

The terms of a Title X loan are designed to encourage development. The interest rate varies with the market but usually remains on the lower side of competitive rates. In addition to the regular cost of interest, an additional charge of one-half of 1% is added to interest for the FHA mortgage insurance fund. After the initial closing, no payments are made for 3 years. The developer agrees to complete his work within the first 2 of those 3 years. The initial sales of lots and parcels from the complete tract occurs during the third year. Equal monthly repayments on the Title X loan begin at the end of the third year, and are so scheduled as to amortize the loan during a seven-year period ending at the end of the tenth year. There is no penalty if the loan is paid off sooner. Parcels of land developed under the Title X loan can be released from the mortgage by paying a proportionate part of the principal. The release clause of the mortgage stipulates that the price of the land sold must be at least 110% of the pro rata mortgage amount. For example, if 10% of the land is sold, that 10% can be released from the mortgage by paying $1.10 \times 10\% = 11\%$ of the mortgage principal. In that way, when 90.9% of the land is sold the entire loan would be repaid in full ($1.10 \times 90.9\% = 100\%$). The terms of the mortgage also permit subordination to a construction loan to provide better security for the construction lender.

8.12.   Title X Example

An example of the potential of Title X is given in the following case. A developer obtained an option to purchase 465 acres of land by paying the owner an initial $5000 and promising to pay an additional $1,000,000 balance if and when the Title X loan was closed. He then hired an architect to design the land development for an initial fee of $1000, with an additional balance to be paid if FHA approval were forthcoming. The design proposed that 90 acres be used for a regional shopping center (the site was near a large metropolitan area), 120 acres developed with multiple-family buildings, 120 acres devoted to town houses, and the remaining 135 acres reserved for single-family homes. All lots were laid out along well-planned streets. With the aid of a mortgage broker, the FHA was persuaded that the design, if carried out, would raise the value of the raw land to $4,000,000.

Therefore, at the initial closing, the developer got the 50% of the raw land value provided for under Title X. In this case that was $2,000,000. (The second million was borrowed money, and thus neither income nor capital gains, so it was not taxable. However it must eventually be repaid.) The second million was used to pay the 10% of the cost of development not covered by the loan. However, as parcels of land in the development were sold, the developer recovered this 10% and more.

The developer then built a sewage treatment plant on land adjacent to the 465 acres, using some Title X money. He then recovered the cost of the sewage plant from tie-in charges levied on individual owners who had bought plots from him on the 465 acres and needed sewer service. These funds, of course, were used to repay the Title X loan on the plant. However, during the course of operation of the sewer system he made a reasonable profit, which was relatively tax free due to the depreciation allowance on the plant. Ultimately he deeded the sewer system to the public and was entitled to a tax deduction in that regard.

Due to his knowledge of available programs, this energetic and enterprising developer was richly rewarded for building a quality product at a reasonable price by providing homes and places to play, shop, and work for the citizens of the area.

9

Front Money:
Where to Find It, How to Get It

9.1. Financing for a Crucial Period

This chapter tells how to finance a vital phase of the project: the design, preparation, and construction phase. This phase begins with the signing of the commitment for the permanent mortgage and terminates when the project is completed and accepted for release of the permanent mortgage financing. The developer is especially vulnerable during this phase because of costs resulting from a long list of necessary steps toward successful start-up of construction, while no income can be expected except funds borrowed by the developer to cover these costs.

9.2. Front Money

Assuming the commitments for both the permanent loan and for the construction loan have been obtained, before construction can start funds must be found to pay for a number of important cost items. During the design and preparation phase money will be needed to make a good-faith deposit on the loan; to pay legal, architectural, design and surveying fees; to meet various other closing costs. During the construction phase money is needed to pay for the building permit and for the bonding fee if one is required; and to order materials and rent equipment and move onto the job site in preparation for the actual start of construction and to make payroll until the first job draw is received. All of this is in addition to any expenditures required to secure and prepare the land, such as demolition or clearing costs. All this cash is called *front money* because it is needed *before* the builder receives the first draw.

9.3. Sources of Front Money

Like every other monetary need of the project, front money can either come from the developer's own capital or it can be borrowed. The two basic reasons in favor of the borrowing alternative are (1) supply: one's own capital is usually quite limited, while the supply of money to lend is quite large, (2) profit: developers and other entrepreneurs usually make more profit using borrowed funds than by using their own because of the leveraging effect. There are lenders that will advance front money as a regular commercial loan to the developer, provided that there is a firm commitment from a mortgage lender as well as another commitment from a construction lender. Since the risk is low for a bank that lends front money, the interest rate should compare favorably with normal short-term commercial loans. The cash from the front-money loan will be needed at the time of initial closing, so it must be arranged well in advance of that date.

9.4. Cost of Front Money

The borrower will pay interest on a front-money loan only for the time that the cash is actually held. (There may also be points or a discount or other fixed front-end charges involved, but the amount of these charges should not be large.) If the interest rate is, for example, 12%, and the amount borrowed is $100,000, at the end of 1 month the borrower will owe $1/12 \times 0.12 \times \$100,000 = \$1000$ in interest in addition to the principal amount. The rate usually is stated on a yearly basis, but the interest is charged monthly, so the actual monthly interest rate is 1/12 of the stated annual rate. Interest that is allowed to accumulate from month to month (compound interest) becomes part of the principal owed and thus earns interest on the interest. If the interest is not paid at the end of the first month, the interest owed for the second month is $1/12 \times 12\% \times \$101,000 = \$1010$.

9.5. The Construction Loan

Where does the developer or contractor find money to pay the bills while construction is under way? Not from the permanent mortgage, since it becomes apparent from a review of the commitment form that the permanent mortgage money will not be released to the borrower until after construction is completed. The permanent lender lends mortgage money only on completed projects. However, another source of funds, called a construction loan, may be obtained to pay the contractors' bills and payroll during the construction process. With a firm permanent-loan commitment in hand, construction loans usually are not difficult to obtain. In fact, the mortgage broker should be able to direct the borrower to at least one commercial bank that makes construction loans, and usually the broker will not charge a fee for this service.

Successful funding of a construction project hinges on obtaining a commitment for a permanent loan. Once a commitment for a long-term mortgage is obtained, all the other pieces of the financial puzzle usually fall easily into place. With the assurance that money will be forthcoming from a mortgage lender upon completion of the construction, it is a relatively simple matter to obtain a construction loan and any other financing that might become necessary.

Although construction loans sometimes come from savings and loan institutions, on larger projects a common source is either a commercial bank or an REIT. For an FHA-insured job the permanent loan and the construction money are often issued by the same bank.

9.6. Construction Loan Application

The developer needs a construction loan as soon as the following steps are complete:

1. A commitment for a mortgage loan has been obtained.
2. A commitment (if needed) for gap financing for the gap amount between the floor and the ceiling of the mortgage loan has been obtained.

After obtaining the gap commitment, the developer approaches a construction lender and gets a construction loan commitment for the maximum amount possible.

The process of obtaining a construction loan is not complex. The developer calls at a selected bank and asks for the loan officer of the bank. The loan officer is given copies of the commitment from an acceptable permanent lender and, if needed, a commitment from an acceptable gap financer. Accompanying these papers are an estimate of actual construction cost, plus a financial statement of the developer's company and personal financial statements. The loan officer of the bank will also want to know what financial reserves are available to meet contingencies, should they occur. In other words, the bank wants assurance that the whole project is going to be completed successfully. They want to be absolutely certain that the amount of money they lend will be enough to complete the project when added to the resources of the borrower. The bank has no desire to be forced to foreclose on a defaulted loan and take over a partially finished project. The loan officer does not want to have a builder come back some time later and say, "I'm sorry, but I underestimated my costs and need more money." In such a situation the bank almost has to lend more money to protect the substantial investment they already have in the project. The bank will seldom get their money back out of the permanent loan unless the whole project is completed, because it is the permanent loan on the completed project that pays off the construction loan. Therefore, the loan officer will require that the developer sum up all project costs to begin with and show exactly how these costs are going to be met.

If the construction lender is satisfied that the cost estimates are accurate; that the borrower is of sound character, well-motivated and competent to build the project and repay the loan; and the project qualifies in all other respects, the construction lender will issue a commitment.

Interest is paid only on the amount of construction loan money actually borrowed. Interest starts accumulating at the time of each draw. The very day that the developer actually receives the money from the bank interest begins to accrue on that money. Each draw accrues the interest assigned to it. Upon completion of construction the full amount of interest due is the sum of all the interest obligations on all of the draws.

When construction is completed to the satisfaction of the loan officer, the last draw is issued, bringing the total loaned up to the total amount of the construction loan less any interest owed. The bank will withhold that last portion of the loan that equals the interest owed on previous draws. For example, suppose that the construction loan commitment were for $100,000 and the total interest owed to the bank at that point is $5000. Then the last draw will bring the total amount received from the bank up to $95,000 (calculated as $100,000 − $5000 = $95,000). However, although the builder has received only $95,000 in cash, the full $100,000 must be repaid since $5000 interest is due on the note in addition to the amount borrowed. If the house is built on contract for a specific owner, then customarily the draws are debited to the owner. Typically each draw from the bank is made out both to the owner and to the builder. The owner receives it first, then endorses it over to the builder. It is the owner who is obliged to repay the loan to the bank, and for small projects, the construction loan becomes the permanent mortgage without any further paperwork. At the conclusion of construction, the owner begins repaying the $100,000 loan by making uniform monthly payments to the bank based on the interest rate and duration of the loan, until the entire $100,000 is repaid. The construction interest is part of the cost of construction, and is included in the amount borrowed under the permanent mortgage.

9.7. The Effect of Discount Points

Even though the lender's risk on a construction loan is reduced by the commitment for the mortgage loan, some risk still remains in a short-term building loan. Because of the need for compensation to offset this risk, the construction lender normally asks for a discount of perhaps 2 points or more. This discount either is deducted from the loan before any money is disbursed or added to the balance to be repaid, and effectively raises the actual interest rate paid. Consider an example: Suppose that the bank makes a loan of $10,000 at 12% nominal annual interest, with actual interest being 1% per month simple interest paid monthly. By the end of the year the borrower will have paid the bank $100 per month interest for 12 months plus repaying the original $10,000 for a total payment of $11,200. If the money is borrowed for only 6 months, the repayment will be only $10,600 since the interest on $10,000 for 6 months is just half the interest for 1 year. If at the time the loan was made the borrower paid the bank a non-refundable front-end fee of 4 points, or 4% of $10,000, or $400, that fee would be income for the bank. The bank has received $400 from the borrower at the time the loan was paid out. By the end of 6 months the bank will have received another $600 in interest from the borrower (at $100 per month) in addition to the repayment of the $10,000 loan. Thus, for letting the borrower use its $10,000 for 6 months, the bank has received a total income of $1000 made up of the first $400 charged as points on the loan and the last $600 charged as interest. Therefore, in truth, the bank has received $1000 in interest over the 6-month period on the $10,000 loan. If the bank makes the same arrangement with the next borrower for the next 6 months, it will have received a total of $2000 for the use of $10,000 for 1 year, so that the bank will actually be earning 20% on its money and not 12%. As will be shown later, the true effective interest is even higher. The 4-point discounts paid on the 6-month example just given are equivalent to 8% additional interest on the $10,000 loan because the points are paid at the beginning of the 6-month loan period. The ratio of points to percentage of interest decreases with increases in repayment period.

The nominal interest rate on the construction loan is usually about the same as the interest rate on the permanent loan, but the discount points increase the true interest. Thus, a 12% nominal interest rate with a 4-point discount will produce a yield of perhaps 20% or more for the bank, depending on the length of time the money is loaned out.

9.8. Commercial Loans

The construction loan is borrowed either by the developer/owner or by the contractor. The developer/owner can usually borrow at a slightly lower interest rate, but some developer/owners prefer the contractor to include the extra cost of construction financing in the contract bid. When the contractor provides the construction financing, the developer/owner pays only one lump sum contract amount upon completion of the job. The construction loan is generally placed with a commercial bank because it is a short-term business loan. Whether borrowed by the developer/owner or the contractor, a short-term loan usually is one which is due somewhere between 90 days and 3 years, which is the maximum time limit of most commercial-bank construction loans. It is a business loan because the money is in essence loaned to a builder so that the business can operate and hopefully make a profit. The business of the builder is to put up buildings, and once the buildings are completed and accepted the construction loan can be paid off with the proceeds of the permanent mortgage loan. Provision should be made in the construction loan contract for delays in completion and other unforeseen circumstances. Lenders are often quite understanding when it comes to modifying the terms of a loan contract if necessary. The commercial bank does not want to foreclose on the construction loan and become the owner of the buildings. It has no desire to go into the real estate business. The bank is in the money business. It wants to rent its money at a reasonable rate called interest, and then it wants to get its money back and rent it out again.

9.9. Qualifying for a Commercial Loan

In order for a borrower to get money from a bank, three requirements must be met:

1. The borrower must have a reputation for honesty, integrity, and consistent repayments on loans.
2. The borrower must show that the money will be prudently invested for a productive purpose.
3. The borrower must show that the return from the prudent investment will enable the loan to be repaid on time.

The builder's reputation is established over time by conducting business affairs in an honest and straightforward manner. The builder's prudent investment is in the plans to erect quality buildings at a fair and competitive price. The source of the funds to repay the construction loan is the permanent mortgage loan which will be released upon satisfactory completion of construction.

9.10. Conformity of Commitments

The commitment from the permanent lender to the developer provides a reasonable assurance to the commercial bank that its construction loan will be repaid. In order to eliminate possible conflicts in provisions, the two loans should involve the same terms. Therefore the construction lender should insist that those appropriate paragraphs of the commitment form discussed in Chapter 6 apply also to the construction loan.

9.11.   *Retainage on Construction Draw, Non-Single Family*

Assume now that the front money has been advanced, and construction has started. During the first month of activity no further money will be forthcoming from sources outside the contractor's own funds. At the end of the first month the construction lender will let the developer have a certain number of dollars, the amount depending on the value of the work put in place. Suppose, for example, that $100,000 worth of work has been put in place during the first month. The builder will then submit a request for a draw of $100,000 against the total that was committed as a construction loan. The request will have to be approved by the bank's representative, perhaps an architect appointed by the bank. The bank will then let the builder have the money. However, the bank will not lend full value, but will retain a certain percentage of the request. Generally this percentage will be either 10 or 20%. The retainage will be spelled out in the construction lender's commitment. The reason for a retainage is to provide the lender with some additional leverage to insure satisfactory completion of the job. However, the best guarantees of satisfactory work are always the reputation and integrity of the builder, a well-drawn contract, and a good design.

9.12.   *The Last Draw*

Suppose again that the first draw request were for $100,000 and that the retainage is going to be 10%. In that case the builder will receive only $100,000 less 0.10 × $100,000, or $100,000 − $10,000 = $90,000. If at the end of the second month an additional $150,000 worth of work has been put in place, bringing the total value of work in place up to $250,000, the builder will receive an additional amount equal to 90% of $150,000 or an additional $135,000. At the end of the first month $90,000 was received and at the end of the second month a total of $90,000 + $135,000 = $225,000 was received against $250,000 worth of value put in place. When the entire job is completed, at the time of the last draw, all the retainage will be added to the last request, so that the builder will finally get all the money that was committed to the loan on the project. The bank will not pay the builder any interest on the money that was held out, nor does the builder have to pay the bank any interest on the money not received.

9.13.　Required Capital

During the course of the first month the front money is used to meet all obligations. At the end of the first month, when $90,000 is received, the front-money loan can be paid off, but because of the retainage there will be no extra money to meet the obligations of the second month. This means that the builder has to find money to pay labor, overhead, and materials suppliers. If the net construction draw is not sufficient, the builder must use his own cash or be able to borrow it. Contractors who prefer to use their own money find that they need about $1 in the bank for every $10 in the contract. With this reserve they usually do not have to borrow anything during the course of construction except for the construction loan.

9.14.　Contractor Borrowing on the Construction Contract

The additional funds required by the builder to finance the project above and beyond the construction loan can be borrowed, possibly from a commercial bank as a business loan. The security for the loan is the construction contract itself. The contract between the contractor and the owner stipulates that a certain sum of money will be paid provided a certain service is performed, that is, provided the project is built in accordance with the owner's plans and specifications. From the bank's point of view, it can lend the contractor money so that he can perform that service. This is a satisfactory arrangement. The contract in a sense guarantees payment, in a manner similar to that in which the permanent loan commitment guarantees payment of a construction loan. Since a profit is anticipated on the construction contract, the commercial bank believes that it will be paid back from the proceeds of the final draw.

When the developer and the contractor are the same person, it may seem strange that additional funds can be borrowed, but actually two different operations performed by two different companies should be involved. The business of the developer's company is to create the entire project from concept to successful operation, including the purchase of land, arranging for design, construction, and so forth. The business of the contractor's company is solely building, which constitutes only one part of the entire project. Although a person cannot sign a contract with himself, it is possible for one company to sign a contract with another company. The contractor might have a corporation for building, while as an individual he is one of the limited partners in a development company. Therefore the builder would have a valid contract with the developer and could use that contract to obtain interim financing during construction.

The interim financing goes on from month to month during the course of the job. It may be paid back each month, or it may be in the form of an open note which is renewed at the end of each month, when accrued interest is paid on the principal until the end of the job. For instance suppose that $10,000 is borrowed to meet the labor payroll during the second month. This has nothing to do, really, with the amount of the holdback; it is simply the amount that the contractor needs to meet ongoing expenses. If there had been no holdback, the contractor would have been richer by the amount held back, and would have had that much more operating capital. He might still have borrowed the money in preference to using his own capital, holding on to his money for other purposes. The decision to borrow or to spend is a management decision that involves many factors. Suppose that the contractor elects to borrow. At the end of the second month the note becomes due. He can pay back the $10,000 plus the interest, or he can pay the interest only. If he pays back principal and interest, it is quite likely that he will immediately have to borrow the same amount again in order to meet similar bills during the third month. If the loan is left in place, he will have enough money in the bank from the construction draw to meet his obligations during the third month. The two methods are about the same, except that in some states each time money is borrowed certain small documentary stamp taxes or other fees must be paid on the transaction. At the end of the project the loan should be repaid in full so as to maintain good credit standing for the next project. Bankers recommend that all open notes be cleared up at least once each year. This is good banking policy as well as an item that the bank's auditors watch for.

9.15. Picanos Villa Apartment Project—Construction Loan and Gap Financing

Remember that the mortgage loan commitment for the Picanos Villa Apartment project was for either a ceiling of $2,883,000 (if 81% rental occupancy is attained), or if rental occupancy is lower than 81%, a floor of 80% of the ceiling or 0.8 × $2,883,000 = $2,306,400.

To minimize its risk in making construction loans, a commercial bank will conservatively base its construction loan on the lesser floor amount of $2,306,400 unless a commitment to finance the gap is obtained. The *gap* is the difference between the floor and the ceiling mentioned in the long-term mortgage commitment. In this case the gap equals $2,883,000 (ceiling) − $2,306,400 (floor) = $576,600. The developer needs to find a lender who will guarantee to fill in this gap in case the rent rolls do not reach 81% occupancy and the mortgage lender lends the floor amount of $2,306,400 instead of the ceiling of $2,883,000. There are lenders who make special loans designed to fill this need and will issue a stand-by commitment covering the gap. They are willing to back up a builder's firm conviction that his project will be successful and will indeed achieve its required rent roll within the allotted time. However, if the rent roll achievement is not met, and the mortgage lender will fund only the floor of $2,306,400, the stand-by lender must be prepared to lend the other $576,600 making up the gap. The stand-by lender will then have a second mortgage with no specific security. Since this can be a high risk venture, the cost of a stand-by commitment is commensurately high. Often the requirement is a fee of 5 points (5%) on the amount of the gap funding, cash in advance at the time the stand-by commitment is made, and the stand-by commitment must be made before the construction loan commitment is obtained. On a matter of $576,600, 5 points would amount to $0.05 \times \$576,600 = \$28,830$. This $28,830 is not refundable and is paid only as an inducement to the lender to issue a stand-by commitment. Even if the 81% rent roll is achieved and the lender never has to take one dollar out of his bank account, he can keep the entire sum of $28,830. The reason for paying such a large fee is to obtain security for an increase in the construction loan, thus inducing the construction lender to increase his loan by the gap amount, which in this case, would be an increase of $576,600. The developer's payment of $28,830 has increased the construction loan from $2,306,400 to $2,883,000, and there is now enough money to do the job.

This fee of $28,830 should not be regarded lightly since it is a relatively large sum. But it becomes a necessary expenditure if there are no other less expensive ways of financing the gap. To the developer there is a vast difference between coming up with $28,830 and providing the gap amount of $576,600 out of scarce company capital.

9.16. Cost of Gap Financing

After the construction is completed, the permanent loan will be funded. By then suppose that only 50% of the rent roll is leased instead of the required 81%. In that case the permanent loan will be $2,306,400, and the stand-by gap lender will provide an additional $576,600.

Interest must be paid on all borrowed money. The gap financing may carry an interest rate of 2½% or more higher than the rate on the permanent loan. If the permanent lender has an interest rate of 12%, the gap loan may be at 14½% or more.

In accordance with the commitment agreement, if during the month or two immediately after completion of construction the 81% rental roll requirement is met, the permanent lender will bring the permanent loan up by an additional $576,600 from the floor amount of $2,306,400 to the ceiling of $2,883,000. The borrower can use the additional $576,600 to pay off the gap financer who gets back his original $576,600 plus the $28,830 paid before a construction loan was made, plus any interest due for the use of the $576,600 during the month or two that the borrower actually held it.

9.17.   Stand-By Commitment on the Gap

Looking at the sum of all project costs up to this point, the developer must determine if the construction loan will be sufficient to get the project built. If not, where can additional funds be obtained? If the construction loan is going to be at the floor level of $2,306,400 and the actual construction cost will be $3,005,330 including construction loan interest and all other costs, there remains a difference of $698,930 that has to be obtained if the project is to be completed. The money might be borrowed from the construction lender provided that the lender receives an adequate guarantee of repayment. Banks insist on security, since inadequate security could soon lead to bank failures as defaulted loans become uncollectible and the bank loses its assets along with its depositors' confidence. A bank acts as an agent holding the money of a community of depositors and has an obligation to safeguard its deposits. Since quite a bit of the permanent mortgage money will be used for land, land development, fees, interest, and closing costs, the construction lender will probably lend only about 80% of the permanent mortgage amount. The amount of the mortgage loan ceiling is $2,883,000, while the mortgage loan floor is $2,306,400. Assuming a gap commitment has been obtained, the construction loan would be equal to the mortgage ceiling of $2,883,000. Without gap financing the construction loan would not exceed the mortgage floor of $2,306,400. Now suppose that out of a total construction cost of $3,005,330 the builder's fee and overhead is 15%, or $450,799, leaving an actual out-of-pocket builder's construction cost of $2,544,531. If the construction loan is for only $2,306,400, there remains a differential of $2,554,531 − $2,306,400 = $248,131 between the construction loan and the actual construction cost. Where is this money to come from?

It is not uncommon for the general contractor to do 20% to 30% of the work, with the rest done by subcontractors. Suppose in this case that the general contractor does 25%, or $638,633 worth of work, and that the subcontractors do 75% or $1,915,898 worth. The total cost of all the subcontractors' work includes the total overhead and profit of each of them. If 10% retainage is held back on each of the subcontractors, then 10% of $1,915,898 yields a total of $191,590 held back until the end of the job. This helps close the gap on the differential of $248,131 that existed between the construction loan and the total cost of construction. Remember that the total cost of construction includes legal fees, closing costs, and so on. It is made up of all costs involved in putting the *brick and mortar* in place and of the building and improvements, exclusive of land costs. The cost data are summarized in Figure 9.1. The construction loan is slightly greater than the amount of money needed to complete construction. The differential would be used to meet contingency costs that inevitably occur.

The general contractor and, in turn the subcontractors, will all be paid

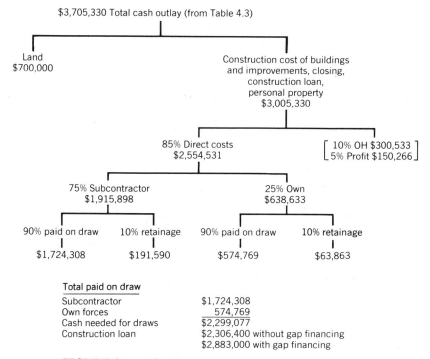

FIGURE 9.1. Distribution tree for costs of Picanos Villa.

in full when the final long-term mortgage is funded at the floor or, preferably, ceiling as the case may be.

9.18. Fairness of Holdback on Subcontractors

If the bank is going to withhold 10% of each draw, then most reasonable subcontractors will also agree to a 10% holdback. It would not be fair for the subcontractor to demand payment in full when the bank is only disbursing 90 cents on a dollar's worth of completed work. The subcontractor can anticipate receiving the balance due at the completion of construction, according to customary practice. Payment practices such as this should be spelled out in the contract so that both parties know what to expect.

Basically the costs of financing the project will fall on the owner, since experienced contractors and subcontractors simply add their financing costs to the bid price. The owner is wise to recognize this fact and make provisions openly to finance the work since the owner usually can borrow money at a lower interest rate anyway.

Any money that is borrowed from a bank must be paid back with interest. To reduce the cost of interest some contractors employ a system that amounts to forced "borrowing" from their subcontractors and suppliers. An unscrupulous contractor will accomplish this by not paying bills when they are due, but stalling until some future date. If a bill is not paid, then money is owed on that bill. If money is owed, then it has in a sense been borrowed from an involuntary lender. A subcontractor who has performed a certain amount of work is entitled to payment for it. If the subcontractor does not get paid, in a sense the general contractor is forcing the subcontractor to lend the money in question. Consequently the subcontractor may be so short of capital as to have to borrow from a bank causing an unexpected expense not covered in the subcontractor's bid price. The financial difficulty might become severe enough to impair the subcontractor's ability to make payroll, and this could put the completion of the entire job in jeopardy. The practice of deliberately delaying payment to subcontractors and suppliers is unethical and unfair and should be severely condemned. Subcontractors should insist that their subcontracts include provisions for penalty fees to be charged against the contractor for any payments that are not made within a reasonable time.

9.19.   FHA Eliminates Need for Front Money

When the FHA is involved in a large residential project, no front money is needed. The first draw of the construction loan is made at the time of the initial closing, and amounts to a substantial percentage of the builder's costs up to that date. This is a much more sensible way of doing business because it involves fewer lenders. Furthermore, it presumes that project owners are trustworthy professionals and it treats them accordingly. However, for conventional loans no draw-money is forthcoming from the construction loan until about 1 month after the start of construction, which likely will be more than a month after the initial closing, and also after the developer has incurred a number of heavy expenditures.

9.20.   FHA Loans

The process of obtaining construction money is also simplified when the FHA (which is now a part of the Department of Housing and Urban Development, HUD) is involved. When the loan is to be insured by the FHA, the entire process is about as complex as obtaining a small home loan, even though it may involve $1,000,000 or more. The FHA insurance, which normally costs ½% of the mortgage balance per month, has the sole purpose of guaranteeing that the loan will be repaid to the bank in the eventuality of a default. The guarantee is backed by the full faith and credit of the government of the United States plus a sizeable insurance reserve fund. This insurance is not the type that protects against damage or fire, but is simply insurance against default on the mortgage. The bank thus has a guarantee that the loan will be repaid, and the risk factor is virtually eliminated. The government will actually make the mortgage payments only until it finds a private buyer for the property, but the money will flow to the bank from some source. Because of the governmental guarantee of payment in case of default, the final lender can make progress payments during construction in the same way that a savings and loan bank does for a house. In other words, even for a substantial project, the construction lender and the final lender can be the same bank.

9.21. FHA Requirements

Of course, the borrower has to pay a small premium for this insurance, usually about one-half of 1% of the amount of the mortgage. In addition, the FHA design requirements are much stricter than those of a conventional lender. They call for certain distances between buildings, certain minimal sizes for rooms, certain amounts of storage, and so on, that can prove restrictive. These are all spelled out in the Minimum Property Standards (MPS) of the FHA, both for single-family and multiple-family housing. Another difficulty with FHA financing of apartment complexes is rent control. One would expect such a restriction of subsidized rental programs, but it applies also to nonassisted programs which place no limit on the income of the tenants. Rents in such projects can be raised only after receiving written permission. Approval is usually forthcoming quite rapidly, provided the owner can prove that the increase is needed because costs have gone up since the rent schedule was established. However, an owner often cannot obtain permission for as high a rent as is charged in a project next door, which has units of exactly the same size and exactly the same amenities. This means that the owner's profits are restricted to a level determined by the government. The FHA does not insure ventures that are totally commercial in nature, such as office buildings or motels, but a project that is mostly residential, with a small amount of auxiliary commercial rental space, can qualify for an FHA loan. Nursing homes and related facilities can also qualify.

9.22. FHA Loan Benefits

The two obstacles of the small premium paid for the insurance and the design requirements, are almost trivial in comparison to the benefits derived from an FHA loan. To begin with, the interest rate is lower with an FHA loan than with a conventional mortgage. Usually the initial discount points on the mortgage are higher because the bank may want a higher yield on its loan than the maximum interest it is permitted to charge on an FHA loan. Theoretically the bank should be satisfied with a lower interest rate when the risk factor is reduced by use of the FHA insurance, but the banks often conclude that the interest rate established by the government for FHA loans is just too low. However, the true interest rate, even considering the effect of a higher discount, normally is lower than that for a conventional mortgage, and the period of time for repayment usually is longer, leading to lower monthly payments for an FHA loan.

A second benefit of FHA mortgages involves leverage. Whereas a conventional lender will give only 65 to 80% of the value of a completed project, an FHA-backed loan may cover as high as 90 to 97% of the value, depending on the project and its sponsor (i.e., depending on whether or not the project is built for profit, whether or not the subsequent rental is subsidized, and so forth). The FHA was established to increase the quantity and quality of housing in the nation, and it does an excellent job in that role. Anyone who qualifies can have a successful project with a minimal investment if FHA rules are followed. More information on various FHA programs is given in the Appendix. The FHA also insures single-family home loans, and the loan procedures are so much like that of a savings and loan bank that they will not be discussed here, the major difference being in the amount of the loan.

9.23. Conventional Mortgage Insurance

To avoid governmental standards and regulations, and yet secure a 90% loan on a single-family residence, it is possible to obtain conventional mortgage insurance directly through a bank that is making the loan. The bank acts only as an agent on the insurance, since the issuing company is usually located in some other city. Usually the insurance is obtainable upon request by any qualified buyer.

9.24. Single-Family Home Loans

For smaller projects such as single-family homes, the entire procedure is greatly simplified. Here there is no need to get three different commitments before construction can start. One commitment is sufficient. In fact, the construction loan and the mortgage loan are typically obtained from the same bank. The initial discount is the only fee paid to the bank.

9.25. Single-Family Draws

As construction work progresses on the project, construction loan money is released to the builder from time to time, as will be discussed in Chapter 10. These progress payments are called *draws*. The bank usually has a formula as to what percentage of the construction loan will be forthcoming for each stage of completion. The definitions of "completion" for each of the stages is explained as well. For example, when the foundation is finished, a certain portion of the loan is released. After the house is dried in, an additional amount is released, and so on.

9.26. Speculative Housing

A speculative house is a house constructed without any particular buyer in mind, but with the hope of selling it to some customer upon or soon after completion. For speculative housing the bank will frequently extend a construction loan beyond the construction completion date and accept payments of interest only until such time as the house is sold by the builder. There is a time limit on this grace period (generally about 1 year), and the amount of the construction loan will be considerably less than the market value of the house. The monthly interest payments made by the builder will be only on the actual construction loan. Some banks, in attempting to modernize their services, provide a complete financial package. The construction loan becomes a part of the final mortgage loan, all issued by the same bank. There is no floor, no ceiling, no gap, no rental achievements. It becomes a much simpler process for the developer.

9.27. Developer's Temptations

The lender in a construction loan may still incur some risk even when relying on the commitment for a permanent loan. This risk lies mostly in the character, ability, and experience (or lack thereof) of the developer. A developer in this sense is one who formulates a project from concept to completion including financing, design, and construction. The developer puts together the entire package and acts as captain of a team made up of the architect/ engineer designer, the attorney, the owner of the land, the banker, the builder, and so on. This team works together on land acquisition, design, financing, construction, and management. The developer must be capable of handling large sums of money wisely, making provisions to repay the loans on time and avoid financial difficulties.

Occasionally, during the time period from the beginning of construction until the rent-roll requirement is achieved, an inexperienced or unethical developer might encounter financial difficulties on other ventures outside of the project under consideration. The developer could even be tempted to use the lender's money to pay debts and obligations from these other ventures, even though this behavior is unethical and illegal. Developers and builders seldom fail financially because of the project currently under construction. If it does occur, financial disaster usually results from debts incurred on one or two preceding jobs. If the builder still owes debts on the past one or two projects, there is strong temptation to misuse the money that is advanced to pay bills on the present job and instead to use it to pay off past debts incurred on previous jobs. If money for the current project is diverted elsewhere then the current project may not be completed, and the repayment of the construction loan will be in jeopardy. To salvage its loan, the bank may then be forced to take over the construction job and complete the job itself.

9.28. Summary

The financing of a real estate project begins with obtaining a commitment for a mortgage loan. With this in hand, a commercial bank may be induced to give a commitment to make a construction loan. Both the long-term loan and the construction loan contain the same requirements and stipulations, and the borrower will usually pay a fee or discount of 1 or 2 points (1 point is 1% of the total loan) in addition to any money borrowed on each loan. This discount serves to increase the effective interest rate.

The mortgage commitment will have a floor, or minimum loan, and a ceiling, or maximum amount, which will be forthcoming only if the rent roll indicates a required percentage level of occupancy. The amount of the construction loan is based on the floor. The difference between the floor and the ceiling is the gap. To increase the amount of the construction loan, a stand-by commitment that covers the gap may be needed.

During the course of construction, the funds from the construction loan will be paid out periodically in amounts based on the value of work completed: monthly on large projects, but as each stage is completed on single-family homes. When the lender holds back a percentage of each draw, the builder should retain at least that same percentage from the draws made on him by his subcontractors. If between draws the builder does not have the funds to meet payroll, usually a short-term business loan can be arranged.

Before any construction starts, a substantial sum of money is needed for initial closing costs and other fees. This front money can also be borrowed if a commitment for the permanent loan has been secured.

10

How Good Draw Schedules
Can Keep You Out of the Red

10.1. The Builder's Goal

In this chapter assume that all financing has been arranged and construction has started. It is now the builder's goal to get enough money from the bank to completely cover all expenses.

10.2. Residential Draws

When building a house it is extremely difficult to cover all costs with the money borrowed from the bank, due to the method by which banks disburse funds. Payments are made according to an arbitrary schedule set forth by the bank. A typical example is shown in Figure 10.1.

10.3. Amount of Each Residential Draw

Notice that the money is not advanced in equal amounts, nor is it paid out monthly. Rather, the four payments are made according to the stage of work completed, regardless of how long it takes to accomplish that work. In the schedule shown, the first draw of 15% of the total construction loan will be paid when the foundations are complete, which can occur as early as a few days after the initial closing. Since site clearing, layout, construction of footings, and rough plumbing may have a total cost of less than 15% of the cost of the house, the builder has a chance at this point to recoup a part of the previous expenditures on land, design, and closing costs. Al-

THE CHARTER COMPANY

CONSTRUCTION LOAN PAYMENT SCHEDULE

Builder (Mortgagor):

TOTAL AMOUNT OF LOAN TO BE ADVANCED $_____

CMC CONSTRUCTION LOAN NO._____

LEGAL DESCRIPTION:

1st Advancement (15%) $_____
When ( ) 1st Intermediate FHA VA Compliance
Inspection Report received; or, ( ) some
person designated by the Mortgagee certifies
in writing that Items 1 through 3 are completed

    Lot   Block   Amount

1. Prefoundation work and footing
   poured.
2. Foundation walls and piers,
   including garage.
3. Termite shields, if required,
   and rough plumbing.

2nd Advancement (35%) $_____
When ( ) 2nd FHA VA Compliance Inspection
Report received; or, ( ) some person
designated by the Mortgagee certifies in
writing that Items 4 through 7 are
completed.

    Lot   Block   Amount

4. Subfloor laid or concrete slab
   poured.
5. Framed and dried in (including
   garage), exterior wall and
   sheathing or wood siding and
   felt or concrete blocks, exterior
   window and door frames, interior
   framing and ceiling joist, rough
   hardware.
6. Plumbing complete, including tub.
7. Rough electric complete.

3rd Advancement (30%) $_____
When ( ) some person designated by the
Mortgagee certifies in writing that
Items 8 through 16 are completed.

    Lot   Block   Amount

8. Fireplace and chimney (if applicable)
9. Lathing, rough plaster and finish
   plaster, or sheet rock installed.
10. Exterior wall covering complete
    (brick veneer or stucco).
11. Paint priming, exterior.
12. Sash and glazing.
13. Tile wainscoting and flooring.
14. Finish roofing.
15. Interior trimming, including doors
    and kitchen cabinets on site.
16. Duct-work (if applicable) and basic
    heating unit in.

Final Distribution (20%) $_____
When ( ) Final FHA VA Compliance Inspection
Report received; or, ( ) some person designated
by the Mortgagee certifies in writing that
Items 17 through 22 are completed.

    Lot   Block   Amount

17. Exterior trim, screens and garage
    complete.
18. Painting and caulking complete.
19. Floor sanding and finishing and
    linoleum.
20. Electric fixtures installed,
    plumbing fixtures installed,
    including septic tank.
21. Walks, drives, and finish grading
    and clean up all work, inside and
    outside complete, and house and
    garage ready for occupancy.
22. All releases required by Counsel
    of Mortgagee.

FIGURE 10.1.   The charter company construction loan payment schedule.

ternatively, the extra money may be needed to meet payroll or pay for materials during the next phase of construction. In any event, the money should be spent on the project for which it was loaned and not diverted to pay overdue bills on some other job, only to invite financial disaster later on.

10.4. Certification

Figure 10.1 shows that, even though no architect is involved, the bank requires certification of the completion of each stage of work. In the case of an FHA or Veterans Administration (VA) loan, the government agency involved makes its own inspections and reports. If the government is not involved, the loan officer of the bank or a qualified representative makes the inspections.

10.5. Location Description

The words "Lot" and "Block" (Figure 10.1) describe a location in the subdivision and are numbered on the Subdivision Plat on file in the County Clerk's office. A block is usually surrounded by streets, while a lot is a single parcel within the block. It has happened that a house has been built on the wrong lot, a fact which might explain the bank's caution in requiring the land description to be put down for each draw. It is more likely however that several houses are being built simultaneously by the same builder, and the bank wants to make sure that the individual house being funded by this draw is getting the money assigned to it. Otherwise funds might be drawn by mistake from some other house, which could be more expensive and thus entitled to more dollars of draw for the same stage of completion.

10.6. Versatility of the Form

The second draw of 35%, when added to the first draw of 15%, brings the total funding up to one-half of the total loan. At this stage the house is ready for drywall or plaster, as the case may be. The form shown in Figure 10.1 is a general one, and has been prepared so that it can be used for almost any house that is built. For example, the first floor might be either a slab-on grade, or wood framing over a basement or crawl space. Line 4 on the right-hand side considers both methods of building.

The third draw of 30% brings the total funding up to 80%. Notice that item 16 (duct work) would probably have been completed before any drywall was applied. Some banks therefore include this item in the second draw.

The form shown is typical, but each lender may include variations to suit his needs.

10.7.   Lien Release Form

The last part of the funding of the combined construction-permanent loan is made when the house is complete. An example of a form used to satisfy the requirements of item 22 is shown in Figure 10.2. It is advisable to get every required signature just as soon as full payment is made to any particular subcontractor, to avoid undue delay when the entire job is completed. The subcontractor might be difficult to find just when you need him.

10.8.   Value of Efficiency

Since house construction draws are made at the completion of each phase (rather than monthly), the most important single factor in house construction financing is the speed of construction, which in turn depends upon organization and efficiency. If the entire job can be finished in less than 2 months (not an impossible accomplishment with today's technology), it is likely that the builder will be able to pay all labor and material bills on time and without difficulty. But as the time schedule for finishing the job stretches out, the contractor is increasingly likely to encounter financial problems. When subcontractors finish their particular subcontract work they expect to be paid, if not immediately, certainly by the tenth of the following month. If the rest of the work included in the same grouping within that draw is not completed, the contractor will not be entitled to money from the bank for the completed draw and will not be able to pay the subcontractor as expected. For instance, if the tile work and roofing are both included in the third draw, the tile work may be complete, but if the roofing is not installed, the third draw will be delayed, and the tile subcontractor will not be paid. The word will get around, the other subcontractors will be hesitant to work for fear they will not be paid either, and the job will slow down even more. Furthermore, the contractor will lose discounts applicable to timely payments on materials, and the interest paid on the construction loan will increase.

10.9.   Monthly Draws on Commercial Construction

On larger jobs a contruction loan is disbursed monthly by the bank. Each month the contractor makes out a request for payment and gives it to the supervising architect or engineer for approval. Approved copies are then sent to the owners, and upon the owner's approval the bank issues a check

We, the contractor, the following sub-contractors and materialmen hereby acknowledge receipt of payment in full for all work, labor and materials furnished on construction work on property located , in Alachua County, Florida, and more particularly described as follows:

and we release all of our respective rights of lien against said property. Said property belonging to......
.... located at...
in the City of Gainesville, County of Alachua, State of Florida.

Dated at Gainesville, Alachua County, Florida:

DATE	DATE
(...............) ....(SEAL) *Architect*	(...............) ....(SEAL) *Paint*
(...............) ....(SEAL) *General Contractor*	(...............) ....(SEAL) *Painting*
(...............) ....(SEAL) *Brick*	(...............) ....(SEAL) *Plaster Materials (Dry Wall Materials)*
(...............) ....(SEAL) *Bricklaying and Blocklaying*	(...............) ....(SEAL) *Plastering (Application of Dry Wall)*
(...............) ....(SEAL) *Carpentry*	(...............) ....(SEAL) *Plumbing Fixtures*
(...............) ....(SEAL) *Cinder or Concrete Blocks*	(...............) ....(SEAL) *Plumbing*
(...............) ....(SEAL) *Concrete Work*	(...............) ....(SEAL) *Pump*
(...............) ....(SEAL) *Doors*	(...............) ....(SEAL) *Ready Mixed Concrete*
(...............) ....(SEAL) *Electric Fixtures*	(...............) ....(SEAL) *Refrigeration*
(...............) ....(SEAL) *Electric Wiring*	(...............) ....(SEAL) *Roads (Grading, Paving, Clearing)*

FIGURE 10.2. Lien release form.

(............) _____(SEAL)
    Excavation

(............) _____(SEAL)
    Flooring (Terrazzo, etc.)

(............) _____(SEAL)
    Floor Scraping

(............) _____(SEAL)
    Glass (Windows)

(............) _____(SEAL)
    Glass Work

(............) _____(SEAL)
    Gutters and Downspouts

(............) _____(SEAL)
    Hardware

(............) _____(SEAL)
    Heating Equipment (Air-Cond.)

(............) _____(SEAL)
    Iron and Steel

(............) _____(SEAL)
    Landscaping

(............) _____(SEAL)
    Lime and Cement

(............) _____(SEAL)
    Linoleum

(............) _____(SEAL)
    Lumber (Studs, Rafters, Joists, etc.)

(............) _____(SEAL)
    Lumber (Trim)

(............) _____(SEAL)
    Materials for Insulation against
    Weather Conditions

(............) _____(SEAL)
    Millwork (Including Cabinets)

(............) _____(SEAL)
    Oil Burner

(............) _____(SEAL)
    Built in Oven-Range-Hood

(............) _____(SEAL)

(............) _____(SEAL)

(............) _____(SEAL)
    Roofing Materials

(............) _____(SEAL)
    Roofing Work

(............) _____(SEAL)
    Sand and Gravel

(............) _____(SEAL)
    Screens

(............) _____(SEAL)
    Septic Tank

(............) _____(SEAL)
    Shades

(............) _____(SEAL)
    Sheet Metal Work

(............) _____(SEAL)
    Stone

(............) _____(SEAL)
    Stone Work

(............) _____(SEAL)
    Stucco

(............) _____(SEAL)
    Termite Treatment

(............) _____(SEAL)
    Tile and Marble

(............) _____(SEAL)
    Trusses

(............) _____(SEAL)
    Wall Paper

(............) _____(SEAL)
    Wall Papering

(............) _____(SEAL)
    Weather Strips

(............) _____(SEAL)
    Well

(............) _____(SEAL)

(............) _____(SEAL)

(............) _____(SEAL)

FFM–84 · IV 18 60

FIGURE 10.2 *(Continued)*

140

for the designated amount and debits the amount of the loan to the owner's account. All requests are based on a schedule of values which are prepared by the builder at the request of the owner before starting any work. The schedule of values can be either straightforward or front-end loaded (see Section 10.16), but in either case, it is based on the contractor's estimate and approved by the design professional as well as by a bank representative familiar with construction.

10.10. The Contractor's Estimate of Value for Payment Purposes

A contractor's estimate of the value of work completed for payment purposes is made in the usual way: taking off labor and materials, running the extensions, adding payroll tax and sales tax as applicable, and listing subcontractor bids. Finally, the several sheets involved are brought forward into a summary estimate of costs, as shown in Table 10.1. Notice that labor and materials are not separated herein, and that all applicable taxes have already been included. However, many other items have not been included, such as administrative expense (office overhead), general supervision, temporary utilities, miscellaneous small tools, trucking, and cleanup. Nor has the cost of bonding been listed, if it is required. Finally, no mention is made of the builder's fee or profit. This summary estimate is solely a worksheet, to be retained in the contractor's office and not shown to anyone. It serves as a basis for the draw schedules.

10.11. Draw Schedules

There are, as previously stated, two ways of presenting the first draw schedule, which will be used in the same form for each succeeding draw. Consider the first way, which is quite straightforward. In Table 10.2 notice that the various items that were omitted in the estimate summary of Table 10.1 are now included. In addition, the costs of land, professional fees, and construction interest are also shown; but no mention is made of the builder's fee or of the developer's fee, although these might also be included if they were needed to make the total equal, or exceed, the construction loan.

10.12. Required Payments

In the example of Table 10.2 the total cost (minus profit) of $3,555,064 is greater than the mortgage ceiling of $2,883,000, which in turn could be greater than the construction loan because the construction loan could be as low as the floor of the mortgage, or as high as the ceiling. Suppose that the floor is $2,306,400. Then the construction loan would also be

TABLE 10.1 Estimate Summary: Picanos Villa, Urbane City, Anystate

1. Site work

Site preparation	$ 14,996
Rough grading	4,083
Water supply	25,437
Sanitary sewer	37,305
Electric service and grounds lighting	13,138
Storm sewer	14,735
Curbs & drainage	9,184
Pavement	45,567
Walks	28,750
Swimming pool	34,076
Landscaping	13,678
	$ 240,949

2. Buildings

Excavation	$19,928
Footings, concrete & steel[a]	20,726
Masonry	155,025
Waterproofing	1,928
Concrete floors & cement work[a]	305,542
Carpet	94,661
Rough carpentry	83,130
Millwork[b]	29,289
Windows[b]	52,315
Doors[b]	63,129
Metal work	2,990
Drywall	136,485
Insulation[c]	22,944
Roofing[c]	31,490
Sheet metal[c]	8,459
Painting	105,931
Finish hardware[b]	27,252
Tile	135,692
Kitchen cabinets	66,406
Medicine cabinets[b]	3,401
Plumbing	171,758
Heat & air conditioning	214,408
Electric	230,334
Miscellaneous	19,369
	$2,002,792

3. Personal property

Ranges	$26,586
Refrigerators	40,374
	$66,960
Total cost	$2,310,701

[a]To be included in concrete work.
[b]To be included in finish carpentry.
[c]To be included in roofing.

TABLE 10.2   Summary of Costs for Picanos Villa, Urbane City, Anystate

Item	Value	Percentage Completed	Value Completed	Balance
Land	$700,000	100	$700,000	0
Closing, loan brokerage, tax during construction	107,490	72	77,393	30,097
Construction loan interest	136,340	0	0	136,340
Overhead & bond	300,533	22	66,117	234,416
Site work	118,878	0	0	118,878
Walks & paving	74,317			74,317
Swim pool	34,076			34,076
Landscaping	13,678			13,678
Excavation	19,928			19,928
Concrete	326,268			326,268
Masonry	155,025			155,025
Waterproofing	1,928			1,928
Rough carpentry	83,130			83,130
Finish carpentry	175,386			175,386
Metalwork	2,990			2,990
Drywall	136,485			136,485
Roofing	62,893			62,893
Painting	105,931			105,931
Tile	135,692			135,692
Carpet	94,861			94,861
Kitchen cabinets	66,406			66,406
Plumbing	171,758			171,758
Heat & air conditioning	214,408			214,408
Electric	230,334			230,334
Miscellaneous	19,369			19,369
Ranges & refrigerators	66,960	0	0	66,960
	$3,555,064	23.7%	$843,510	$2,711,554

Costs This Month $843,510

$2,306,400, unless a stand-by commitment is obtained for the gap. If there is no stand-by commitment, there is a difference of $1,248,664 between the total cost of $3,555,064 and the construction loan of $2,306,400 (see Summary of Calculations, Table 10.4). In order to protect itself, the bank that is providing the construction loan will not give out any money on any draws until the developer shows that he has already spent $1,248,664 on the project. In other words, after the first $1,248,664 worth of value has been paid for, and there remains a balance of $2,306,400 needed to finish the project, then and only then would the construction loan begin to be funded. Draws will be made against the balance of $2,306,400, and that money will be used to finish the job.

Even if a stand-by gap commitment is obtained and the construction loan equals the mortgage ceiling of $2,883,000, there is still a difference between the total project cost (minus profit) of $3,555,064 and the total construction loan of $2,883,000, but this difference would be only $672,064 rather than $1,248,664. Once again, the bank would insist that this amount of $672,064 be spent by the developer before any construction money was disbursed, and, similarly, the bank would lend only toward the balance. Notice in Table 10.2 that actually $843,510 has been spent, as shown in the "Totals" row, under the column headed "Value Completed." All of this $843,510 was spent before any construction took place. The bank might therefore be induced to lend the difference between $843,510 and $672,064, which is $171,446, at the time of the initial closing, but in all likelihood it would refuse.

10.13.   Reasons for Holdback

The bank probably would not lend this money for several reasons. First of all, a construction lender is making a construction loan, not a land-purchase loan. If no work ever starts, the $171,446 will actually have been put into the land alone. Second, up to the time of the final draw, the lender withholds 10% of all value in place. There are several reasons offered for this 10% holdback, as follows: (1) the holdback discourages an unscrupulous contractor from absconding with the money because the profit has already been made and the contractor has lost incentive to complete the job; (2) the 10% holdback is a reserve fund to meet any bills that the contractor has incurred and failed to pay for some reason; (3) the 10% holdback represents the contractor's profit, and a contractor is not entitled to a fee until the job is completed (however, if the builder holds a completion bond, that argument is specious at best); and (4) the 10% holdback covers any disagreements that might arise.

10.14.   Avoiding Difficulties

To simplify and clarify the procedure, it is advisable to have the "total" under the column headed "Value" read the same as the amount of the construction loan. If, for example, the construction loan is to be $2,883,000 (with gap financing), then item 1: "Land" $700,000, could be reduced by $672,064, thus bringing the total of $3,555,064 down to $2,883,000. In that way there would be no arguments or misunderstandings when the draws are presented during the course of construction, and the first draw could be honored without any questions regarding equity. Obviously, if the "total" reads the same as the loan, the developer has already put up his cash equity. The bank will then be able to lend the rest, and the first draw will be honored in full, less the 10% withholding.

10.15. *Alternative Draw Schedule*

Some contractors prefer to show only the items directly related to construction and omit all incidental or related items, such as profit or fee, overhead, and construction interest. Notice that Table 10.3 has only 22 items, whereas Table 10.2 shows 26 items, although both describe the same job. The difference lies in the fact that the first four items of Table 10.2, "Land," "Closing, loan brokerage, tax during construction," "Construction loan interest," and "Overhead & bond," have been omitted from Table 10.3. These four items have the following values:

Land	$ 700,000
Closing, land brokerage, tax during construction	107,490
Construction loan interest	136,340
Overhead and bond	300,533
Subtotal	$1,244,363

Subtracting this subtotal from $3,555,064 leaves a net of $2,310,701 (3,555,064 − 1,244,363 = 2,310,701). But if the construction loan is for $2,883,000, it is necessary to add the difference of $2,883,000 − $2,310,701 = $572,299 to the items of Table 10.3 so that the "total" of this table will equal the construction loan amount (the Table 10.1 total of $2,310,701 plus the added $572,299 will total the construction loan amount of $2,883,000).

10.16. *Front-End Loading*

Before ever getting the first draw, the contractor will have spent a large amount of money. The contractor is required to put out cash not only on the construction itself, but also on such things as the estimate, which is time consuming and expensive, and on setting up the job in the office and in the field, purchasing materials, renting equipment and hiring labor. It is reasonable, therefore, for the contractor to attempt to recapture this cash outlay as rapidly as possible. It is also reasonable to conclude that a contractor needs only a small profit margin on cost items that are fixed and definite, such as subcontractor's bids and the purchase of materials, whereas a larger profit is required on risk items, such as labor performed by the contractor's forces, which might have a cost overrun. For all these reasons

TABLE 10.3   Draw Request for Picanos Villa, Urbane City, Anystate

Item	Value	Percentage Completed	Value Completed	Balance to draw
Site work	$ 160,490	85	$136,420	$ 24,070
Walks & paving	102,760	0	0	102,760
Swim pool	39,590	0	0	39,590
Landscaping	15,890	0	0	15,890
Excavation	27,340	100	27,340	0
Concrete	416,710	20	83,340	333,370
Masonry	214,160	5	10,710	203,450
Waterproofing	2,240	0	0	2,240
Rough carpentry	115,000			115,000
Finish carpentry	242,920			242,920
Metalwork	3,480			3,480
Drywall	158,580			158,580
Roofing	73,090			73,090
Painting	185,300			185,300
Tile	154,570			154,570
Carpet	110,650			110,650
Cabinets	77,160			77,160
Plumbing	180,350			180,350
Heat & air conditioning	249,140			249,140
Electric	247,160			247,160
Miscellaneous	22,520			22,520
Ranges & refrigerators	83,900	0	0	83,900
	$2,883,000	210	$257,810	$2,625,190

$$\begin{aligned}
\text{Less payments to date} &\quad 0 \\
\text{This draw} &\quad \$257,810 \\
\text{Less 10\% retainage} &\quad 25,781 \\
\text{Due this draw} &\quad \$232,029
\end{aligned}$$

the items of Table 10.3 are often *front-end loaded*. The amount of $572,299 is not distributed uniformly among all 22 items, nor is it applied as a uniform percentage equally to each item. If it had been applied as a uniform percentage, it would have been necessary to add

$$\frac{\$\ 572{,}299}{\$2{,}310{,}701} = 0.2477 = 24.77\%$$

to each one of the 22 items. For example, item 5: "Site Work" of Table 10.2 would have been multiplied by 1.2477 × $118,878 = $148,321 to give an item "cost" increase from $118,878 to $148,321. Instead, fixed-cost items are increased by about 5 or 6%, while the contractor's own work is increased by about 25% or 30% in such a way that the "total" comes up to

TABLE 10.4 Summary of Calculations for Picanos Villa, Urbane City, Anystate

Land	$700,000		
Closing, loan brokerage, tax during construction	107,490		
Construction loan interest	136,340		
Overhead and bond	300,533		
	$1,244,363	$1,244,363	
Site work[a]	240,949		
Buildings[a]	2,002,792		
Personal Property[a]	66,960		
	$2,310,701	$2,310,701	
		$3,555,064	
Profit			$3,555,064
			150,266
			$3,705,330
Total cost of project			
Mortgage and construction loan			
Ceiling	$2,883,000		
Floor	2,306,400		
Gap	576,600		
Total cost minus profit	$3,555,064		$3,555,064
Mortgage	−2,306,400(floor)		−2,883,000(ceiling)
Developer's equity	$1,248,664(without gap)		$ 672,064(with gap)

[a]Taken from Table 10.1.

147

$2,883,000, the desired amount. None of the resulting dollar values is exact, and they have all been rounded off to show a zero as the last number.

A certain amount of front-end loading occurs almost intuitively, but this is not intended to imply that this procedure is standard acceptable practice. In fact, many owners attempt to forbid front-end loading in their contract. Table 10.3 is shown only as an illustration of one possible way to front-end load a draw schedule to a reasonable extent. Many other combinations of numbers would be equally effective. It would be wrong, though, to overdo the process to such an extent as to not be able to justify the estimated amounts to the architect and completely lose credibility. For example, it would be wrong to increase "site work" from $118,878 to $691,177 and leave everything else as is. The "total" would be $2,310,701 but the architect would not and should not approve the first draw. Credibility would suffer and suspicion might permeate the whole relationship because of what might be regarded as deceit on the part of the contractor in constructing a poor draw schedule. Comparison of the items in Table 10.3 with those of Table 10.2 shows that the override in site work is

$$\frac{160,490}{118,878} - 1 = 0.35 \text{ or } 35\%$$

whereas the override in plumbing is only

$$\frac{180,346}{171,758} - 1 = 0.05 \text{ or } 5\%$$

The figures under "Value Completed" in Table 10.3 are based on the assumption that this draw request was presented 1 month after construction started.

11

How to Forecast Cash Needs
during Construction

11.1. Planning Expenditures

A well-managed company keeps its initial capital as intact as possible, and in fact tries to accumulate additional capital. This additional capital is leveraged and used to obtain more work and larger jobs, and thus earn more return. However, only a small percentage of total income can be added to company capital since most of the income is spent on payroll and other current expenses. It is essential to plan each job carefully to avoid borrowing from the company capital to meet short-term cash flow needs any more than is absolutely necessary. The same careful thought and planning must go into the expenditure of money as go into the planning and coordination of the actual work of construction, which involves the scheduling of men, materials, and specialty contractors. Thus, before each job starts, the prudent contractor forecasts just how much cash will be needed each month and determines how to obtain that money without utilizing any more company capital than is necessary.

11.2. Bar Charts

Tentative money draw schedules can be estimated before the job starts if enough time is set aside for planning. Most knowledgeable contractors make use of critical path method (CPM) diagrams to establish the length of the job and to speed up the process of construction. From a good CPM diagram it is not at all difficult to draw a *bar chart*. The actual drawing of the bar chart takes very little time, and the chart itself makes it easy to see at a glance just what percentage of each item on the draw schedule should be completed each month.

11.3.   Example of a Bar Chart

Of course it is not necessary to formulate a CPM diagram and then reduce it into a bar chart, but this is a more accurate process than drawing a bar chart based on intuition or experience. In either case the final drawing will resemble Figure 11.1 This example (Figure 11.1) is not complete and is given merely for purpose of illustration. To simplify and clarify the example the construction time period has been cut down to 5 months. Notice that at the end of the first month all the excavation will have been completed, 85% of the site work will have been accomplished, and 20% of the concrete work will be finished; however, it is estimated that only 5% of the masonry will be in place, if all goes according to schedule.

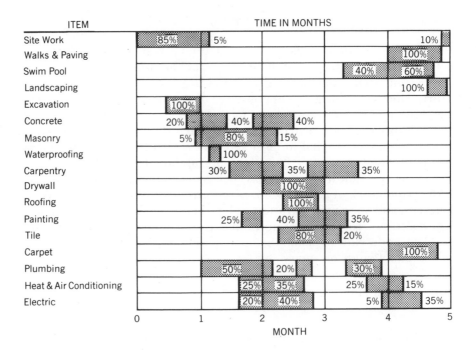

FIGURE 11.1.   Bar chart.

11.4.   Cost Per Month

The estimated percentages of completion can now be applied against the values shown in the Summary of the Costs (Table 10.2) to obtain a good idea of the amount of cash required during each month, as shown in Table 11.1. These are the amounts that will actually have been spent on the job. A similar estimate could be made for each month of the job. All of these several estimates would then be combined into a summary of totals, as shown at the bottom row of Table 11.1 and repeated in Table 11.2.

11.5.   Purpose of Bar Charts

The purpose of the bar charts is to graphically display the anticipated percentages of completion of each item in the job for each month during the duration of the job. These percentages are applied against the dollar values of the items, so as to obtain an estimated total of the dollars required each month as the job progresses. Most of these dollars will come from the monthly draws. The rest will have to come from operating capital, or else the difference will have to be borrowed as front money at the beginning of the job, or as short-term loans as the job progresses, as described previously. The differences can best be seen if the two totals (actual cost vs. draws) are plotted on a graph, as shown in Figure 11.2.

11.6.   S-Curve

The graph has as an ordinate (vertical axis) dollar values, and it has as an abscissa (horizontal axis) the time in days, weeks, or, as shown here, months. When the predicted cash outlay is plotted on the graph, it will be a series of points. The first point will show the amount of money spent before construction starts, which would be $843,510 for the example of Picanos Villa, and will be placed above time zero. The prediction was that at the end of the first month of work $1,135,100 would have been spent, so that point is plotted, and so forth. When all six points of the predicted actual cost have been plotted in Figure 11.2, they can be connected either with straight lines or with a smooth curve. In either event, at this stage the shape of the drawing will look roughly like an elongated or tilted letter "S," and therefore is called the S-curve of cost versus time.

11.7. Plotting Bank Draws

The predicted bank draws can be plotted on the same graph. Each of these draws is based on an estimated draw schedule, which has been calculated

TABLE 11.1 List of Monthly Expenditures by Item for Picanos Villa, Urbane City, Anystate
(Total costs correspond with Table 10.2)

	Expenditures Incurred Before Construction Starts 0	1	2	3	4	5	Total
Land	*100%* 700,000						$700,000
Closing, loan brokerage, tax during construction	*72%* 77,393	*28%* 30,097					107,490
Construction loan interest		*7%* 9,544	*13%* 17,724	*20%* 27,268	*27%* 36,812	*33%* 44,992	136,340
Overhead & bond	*22%* 66,117	*18%* 54,096	*15%* 45,080	*15%* 45,080	*15%* 45,080	*15%* 45,080	300,533
Site work		*85%* 101,046	*5%* 5,944			*10%* 11,888	118,878
Walks & paving						*100%* 74,317	74,317
Swim pool					*40%* 13,630	*60%* 20,446	34,076
Landscaping						*100%* 13,678	13,678
Excavation		*100%* 19,928					19,928
Concrete		*20%* 65,254	*40%* 130,507	*40%* 130,507			326,268
Masonry		*5%* 7,751	*80%* 124,020	*15%* 23,254			155,025

Item							
Waterproofing	100% 1,928						1,928
Rough carpentry / Finish carpentry			30% 77,555	35% 90,481	35% 90,480		83,130 / 175,386
Metalwork			30% 896	35% 1,047	35% 1,047		2,990
Drywall				100% 136,485			136,485
Roofing				100% 62,893			62,893
Painting			25% 26,483	40% 42,372	35% 37,076		105,931
Tile				80% 108,554	20% 27,138		135,692
Carpet					100% 94,861		94,861
Cabinets						100% 66,406	
Plumbing			50% 85,879	20% 34,352	30% 51,527	15%	171,758
Heat & air conditioning			25% 53,602	35% 75,043	25% 53,602	15% 32,161	214,408
Electric			20% 46,067	40% 92,134	5% 11,517	35% 80,616	230,334
Miscellaneous	20% 3,874	20% 3,874	20% 3,874	20% 3,874	20% 3,874	100% 3,874	19,369
Ranges & refrigerators						100% 66,960	66,960
Totals	843,510	291,590	619,559	873,344	466,644	460,417	3,555,064

TABLE 11.2   Estimated Summary of Monthly Expenditures for Picanos Villa, Urbane City, Anystate

Month	Monthly Value	Running Total
Before construction	$843,510	$843,510
1	291,590	1,135,100
2	619,559	1,754,659
3	873,344	2,628,003
4	466,644	3,094,647
5	460,417	3,555,064

using the draw request cost data of Table 10.3 and the schedule illustrated by the bar chart of Figure 11.1. The resulting monthly draw schedule is shown in Table 11.3 and illustrated graphically in Figure 11.2.

11.8.   The First Draw

In a few instances the first draw might occur at the initial closing, that is, at time zero, when construction is just about to start. Such would be the case with certain FHA loans, but in general the first draw does not occur

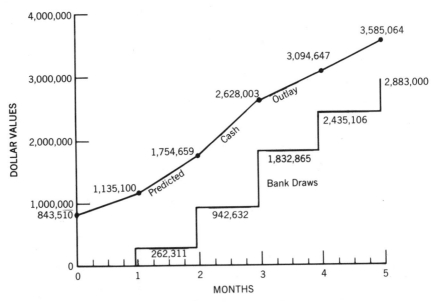

FIGURE 11.2.   S-curve. The time has been shortened to 5 months simply to make the explanation easier. The same principles apply if the job takes much longer.

until after the first month has gone by. In the example of Figure 11.2 it is assumed that conventional financing is used, so the estimated first draw of $262,311 is plotted just above the point that indicates the end of the first month.

11.9. Step Curves

The several points representing the bank draws should not be connected with a curve because the funds are received at monthly intervals and no additional money comes in between these monthly payments. In the case of the actual construction expenditures a curve was drawn as an approximation to the money which is actually spent from day to day on the job, keeping up with the constantly increasing actual cost until the job is completed. However, since the level of each bank draw does not change during any given month, straight horizontal and vertical lines are drawn stepwise to connect the five points shown in Figure 11.2.

11.10. Cash Forecasting

Now the additional cash necessary at the beginning of each month, and also the cash necessary on any given day of the month, can be predicted. At the start of the job $843,510 will have been spent. If front money is borrowed to cover this expenditure, then interest will have to be paid on that front money. During the first month an additional sum of $262,311 will have been spent (as shown in Table 11.3), and the contractor must figure out the best way to finance this amount. A good part of that money is needed to meet the payroll during the first month, since most of the work done during this month is the general contractor's responsibility. The various ways of meeting the payroll are discussed in previous chapters.

11.11. Net Actual Cost

At the end of the first month the first draw of $262,311 will be obtained from the bank (see Table 11.3), but the first draw amount will not be enough to cover all costs up to that date. There will be a net difference of $1,135,100 spent by end of month 1, minus $262,311 drawn at the end of month 1, which equals $872,789. Of course, if the costs of land and closing plus loan and tax costs were omitted, the actual cost at the end of the first month would be reduced by $700,000 (land) + $107,490 (closing, etc.) = $807,490, leaving a net actual cost of only $872,789 − $807,490 = $65,299. (The starting figure of $843,510 shown on the graph in Figure 11.2 is less than $872,789 because it assumes some loan and tax costs will be paid after con-

TABLE 11.3 **List of Monthly *Draws* and Income by Item**

	Income or Investment Before Construction Starts — 0	1	2	3	4	5	Total
Developer's equity	672,064						
Site work		85%	5%			10%	
		136,417	8,024			16,049	160,490
Walks & paving						100%	
						102,760	102,760
Swim pool					40%	60%	
					15,836	23,754	39,590
Landscaping						100%	
						15,890	15,890
Excavation		100%					
		27,340					27,340
Concrete		20%	40%	40%			
		83,342	166,684	166,684			416,710
Masonry		5%	80%	15%			
		10,708	171,328	32,124			214,160
Waterproofing			100%				
			2,240				2,240
Rough carpentry			30%	35%	35%		
			34,500	40,250	40,250		115,000

	Col. 1	Col. 2	Col. 3	Col. 4	Col. 5	Total
Finish carpentry	30% 72,876			35% 85,022	35% 85,022	242,920
Metalwork			100% 3,480			3,480
Drywall			100% 158,580			158,580
Roofing	25% 18,273		40% 29,236	35% 25,581		73,090
Painting			80% 148,240	20% 37,060		185,300
Tile				100% 154,570		154,570
Carpet				100% 110,650		110,650
Cabinets					100% 77,160	77,160
Plumbing	50% 90,175		20% 36,070	30% 54,105		180,350
Heat & air conditioning	25% 62,285		35% 87,199	25% 62,285	15% 37,371	249,140
Electric	20% 49,432		40% 98,864	5% 12,358	35% 86,506	247,160
Miscellaneous	20% 4,504	20% 4,504	20% 4,504	20% 4,504	20% 4,504	22,520
Ranges & refrigerators					100% 83,900	83,900
Totals	262,311	680,321	890,253	602,221	447,894	2,883,000

struction starts.) In fact, the net actual cost of construction only (excluding closing, interest, and overhead) during the first month will be less than $262,311 because the draw schedule has been front-end loaded. The true total costs to the general contractor from the time construction starts until the first draw is received should not be very much more than the first draw if the schedule has been arranged properly (actually $197,853 according to the figure in this example). The actual total of all of the developer's costs, which includes land and fees, will be considerably greater than the draw, however, and this differential will remain large throughout the job.

11.12. Effect of Time Lag

During the construction period the timing of payments of bank draws, as represented in Figure 11.2, may not conform to the dates of actual receipt of the funds by the developer. If the request for draw is not submitted until the last day of the month, the money may not be received until about the tenth day of the following month. Therefore the contractor should forecast an estimate of the amount of work to be accomplished during the last few days of the month, and submit a bill about 3 days before the end of the month. Hopefully the owner's architect or engineer will be able to check the bill, authorize payment, and submit it to the lender bank by the first of the month, in which case the money may be forthcoming as early as the fifth. The time lag will have been shortened as much as possible, but it will still exist and interest costs will continue to mount up. Therefore, the entire set of step lines illustrating the bank draws should be shifted to the right, along the time scale, by about 5 days or one-sixth of 1 month. The gap between draws and actual cost has now become greater, but reflects a more accurate forecast for working out the financial planning for the project.

11.13. Predicting Construction Interest

Another important point is that the predicted actual cost must include interest on the construction loan, particularly if the contractor is also the developer. If the first draw is $262,311, and if the job duration is 5 months, then this money will have been held for 4 months. If the interest rate is 12% (per annum is understood), or 1% per month effective interest, the interest during the 4 months will amount to $262,311 \times (1.01^4 - 1) = $10,650 on the first draw. Each draw will have interest charged to it for the time during which that draw is held. For the example shown in Figure 11.2 the second draw will be held for 3 months, the third draw will be held for 2 months, and the fourth draw will have interest charged to it for only 1 month. Theoretically, the fifth draw will not have any interest charge,

TABLE 11.4 Calculation of Interest Due on the Construction Loan to Cover Construction Draws.

Draw Number	Amount of Draw, P	Number of Months on Loan, n	Interest Only $P[(F/P\ 1\%\ n) - 1]$
1	$262,311	5	0.0510 = $13,378
2	680,321	4	0.0406 = 27,621
3	890,253	3	0.0303 = 26,975
4	602,221	2	0.0201 = 12,105
5	447,894	1	0.0100 = 4,479
	$2,883,000		$84,558

because the permanent financing will take effect at that time. However, there is usually a time lag of 1 month before the permanent financing takes over the construction loan, so that every one of the interest computations for each predicted draw should be increased by 1 month to account for this time lag. The first draw, therefore, will probably be held for a total of 5 months until the construction loan has been paid off by the permanent lender's mortgage. In that case, the total interest due on the first draw of $262,311 will amount to $262,311 $\times$ $(1.01^5 - 1)$ = $13,378, and the total interest is calculated as shown in Table 11.4.

The total construction interest of over $84,000 is equal to more than half of the contractor's profit (assuming the work actually proceeds according to schedule and a profit is earned). But remember that the origination fee of 2 points must also be paid to the construction lender. This 2 points amounts to 0.02 $\times$ $2,883,000 = $57,660 in the example of Figure 11.2. Thus the construction lender will be getting total true interest of $84,588 + $57,660 = $142,218, and this figure must be included in the computation of predicted actual cost.

11.14. Summary

One measure of a well-managed construction firm is an ability to accurately predict how much cash will be needed at regular intervals throughout the job, and then to secure those funds by the time they are actually required. An S-curve can be drawn based upon estimated *actual* costs to the contractor. The draw schedule can then be adjusted to a reasonable extent by a modest amount of front-end loading so that the cash draw requested will come closer to covering the contractor's actual out of pocket costs each month. Front-end loading must not be carried beyond a reasonable point nor beyond any contractually agreed upon limits. The greatest percentage of the estimated surcharge of overhead and profit is placed quite properly on the contractor's own work, and smaller percentages are placed on the various subcontractors' work. All of them are adjusted so as not to violate the terms of the agreement or any standards of reasonableness. These percentages can be applied only one time, and that is when the first schedule of values for completed work is presented to the owner's agent for approval. After the owner's agent approves these values they become fixed. Therefore, to insure a valid schedule of values, sufficient time should be spent in financial preplanning, using a CPM, a bar chart, and the S-curve and step-curve.

2

How to Do a Good
Profitability Study and Find
the After-Tax Rate of Return

12

Basic How-to-Do-It
Time Value of Money Calculations

*Key Expressions in This Chapter**

A = Uniform series of n end-of-period payments or receipts, with interest compounded at rate, i, on the balance in the account at the end of each period.

F = Future value. A single lump sum cash payment occurring in the future at the end of the *last* of n time periods. The future balance of all payments under consideration, together with accrued interest at rate i.

G = Arithmetic gradient increase, or decrease in funds flow at the end of each period (except the first period) for n periods.

P = Present value of all payments under consideration. Present value may be thought of as an equivalent single lump sum cash payment now, at time zero the beginning of the first of n time periods. No interest has accrued. The amount of the present value differs from an equivalent future value by the amount of interest compounded on the present value in the intervening period.

i = Interest rate, assumed to be the annual rate unless otherwise specified. This is the rate at which the balance of the account is charged for use of the funds. Or in nonfinancial problems i is the growth or de-

---

*The words value, worth, sum, amount and payment are used interchangeably. For instance, present value, present worth, present sum, and so forth, all mean the same.

cline of the base number. It is assumed as annual and compounded unless otherwise noted or obvious from the problem.

n = Number of compounding time periods, frequently years, but may be quarters, months, days, minutes or as otherwise specified.

Equations Used in This Chapter

$$F = P(1 + i)^n \tag{1a}$$

$$P = \frac{F}{(1 + i)^n} \tag{1b}$$

$$F = A \frac{(1 + i)^n - 1}{i} \tag{2a}$$

$$A = F \frac{i}{(1 + i)^n - 1} \tag{2b}$$

$$A = P \frac{i(1 + i)^n}{(1 + i)^n - 1} \tag{3a}$$

$$P = A \frac{(1 + i)^n - 1}{i(1 + i)^n} \tag{3b}$$

$$F = \frac{G}{i} \left[\frac{(1 + i)^n - 1}{i} - n \right] \tag{4}$$

$$P = \frac{G}{i} \left[\frac{(1 + i)^n - 1}{i(1 + i)^n} - \frac{n}{(1 + i)^n} \right] \tag{5}$$

$$A = G \left[\frac{1}{i} - \frac{n}{(1 + i)^n - 1} \right] \tag{6}$$

12.1. Cash Flow

As the words imply, cash flow occurs whenever cash or its equivalent "flows" or changes hands from one party to another. When you pay money to the cashier at the lunch counter or at the store, cash flows from you to the cashier. If you pay by check, the check is considered an equivalent to cash so "cash" still flows when you hand over the check. If you pay by credit card you owe the money as soon as you sign the charge slip, but the cash flow does not occur until the credit-card firm receives your check in lieu of cash.

12.2. *Cash Flow Diagram*

Information on cash flow can be communicated quickly and easily by use of a line diagram called a cash flow diagram, a type of graph used extensively because of its simplicity and ease of construction. It consists of two basic parts: the horizontal time line, and the vertical cash flow lines. The horizontal time line is subdivided into n compounding periods, with each compounding period representing whatever unit of time duration is specified for the problem under consideration, such as a year, month, day, and so on. The vertical lines represent cash flow and are placed along the time line at points corresponding to the timing of the cash flow. The vertical lines are not necessarily to scale, although a large cash flow is usually represented by a longer line than a small cash flow.

Example 12.1.   Draw the cash flow diagram from the *borrower's* viewpoint showing a loan of $1000 for 1 year at 10% interest.

Solution: The cash flow diagram from the viewpoint of the borrower is shown in Figure 12.1.

Normally, receipts or incomes are represented by upward pointed arrows since they represent cash flow in and an increase in the balance in the account. Conversely costs, disbursements, or expenditures are usually represented by downward pointed arrows indicating a decrease in account balance due to a cash flow out.

Notice that when the cash flow diagram is drawn for the lender's point of view, the arrows point in the opposite direction than for the borrower. In lending you the $1000, the lender has a cash flow *out* (arrow downward) and upon repayment the lender has a cash flow *in* (arrow upward) as demonstrated in Example 12.2.

Example 12.2.   Draw the cash flow diagram from the *lender's* viewpoint showing a loan of $1000 for 1 year at 10% interest.

Solution: The cash flow diagram is drawn as shown in Figure 12.2.

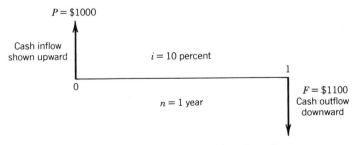

FIGURE 12.1.   Borrower's cash flow line diagram.

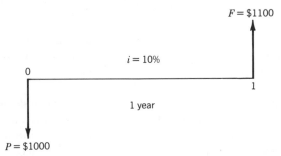

FIGURE 12.2.   Lender's cash flow diagram.

12.3.   Balance-in-the-Account Diagram

Additional information on project financing can be communicated by use of the balance-in-the-account diagram. As the name suggests this graph simply depicts the balance accumulated or left in the account at all points in time. While the cash flow diagram shows all transfers, except interest payments, into or out of the account, the balance-in-the-account diagram shows both the interest payments and the balance before and after the interest payments.

Example 12.3.   Draw the cash flow diagram and corresponding balance-in-the-account diagram for a loan of $1000 for 1 year at 10%, from the borrower's viewpoint.

Solution: Both diagrams are shown in Figure 12.3.

12.4.   Money as a Builder's Tool

To the builder, money is just as important an ingredient of construction as labor, equipment, and material. Money to finance the project is acquired in much the same way as other essential tools—it is either owned or rented (borrowed). When the builder already owns the needed funds they must be withdrawn from the builder's accumulated savings (out of liquid capital funds). These funds then become tied up in the project and can no longer earn interest, so interest is lost or foregone for these funds. The lost interest should be charged as an added cost to the project. When the funds needed to finance the project are rented (borrowed) the lender expects some compensation for doing without the borrowed funds so rent (interest) is charged for every time period (months, years, etc.) that the funds are on loan, and this interest cost also should be charged as an added cost to the project. So

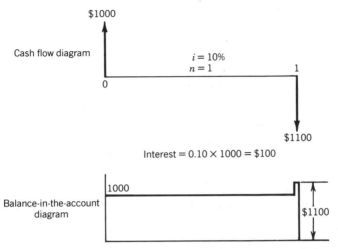

FIGURE 12.3.   Cash flow diagram and corresponding balance-in-the-account diagram.

regardless of whether the project is funded with the owner's money or with money borrowed from a bank, interest costs should be charged against the project for all funds invested therein.

Money has a time value, and the dollar amount of interest paid is proportional to the time the money is on loan. If you borrow $1000 from Bank A for 6 months, and borrow another $1000 from Bank B for 12 months, you expect to pay about twice as many total dollars of interest to Bank B because you held the funds for twice as much time. The dollar amounts of interest payments are very similar to rent payments in that both are charged to enable a nonowner to acquire use of someone else's property for a limited amount of time. Both interest and rent are charged per unit of time, and both are based on the value of the amount loaned or item rented.

12.5.   Interest Defined

More formally, interest may be defined as money paid for the use of borrowed money. Or, in a broad sense, interest can be said to be the return obtainable by the productive investment of capital. The rate of interest is the ratio between the interest chargeable or payable at the end of a period of time and the money owed at the beginning of that period. The rate is usually based on a 1-year period, unless a shorter time is stipulated, such as 1% per month. Thus, if $10 of interest is payable on a debt of $100, the interest rate is $10/$100 = 0.10 per year. This is described as an interest rate of 10%, with the "per year" being understood.

12.6.   Rate of Interest

Sometimes interest is payable more often than once each year, but the interest rate per year is almost always what is meant when the interest rate is quoted, unless specifically stated otherwise. In this way, 0.833% payable monthly, 2½% payable quarterly, or 5% payable semiannually are all described as 10%. If $1000 were deposited in each of four accounts with each account representing one of these four interest rates, there would be a small difference in the account balances at the end of 1 year, due to the differences in compounding. However, the difference between the nominal rate of 10% and the effective rate produced by each of the others is not large enough to affect a decision in the economic analysis of any construction project. (More accurately, the 10% nominal rate reflects a rate of 0.00833 compounded monthly = 10.47% effective annual rate; 0.025 compounded quarterly = 10.38% annual; and 0.05 compounded semiannually = 10.25% annual.)

12.7.   Plans for Paying Back a Loan

Consider the four plans shown in Table 12.1 and illustrated by cash flow diagrams in Figure 12.4 by which a loan of $10,000 is paid back in 10 years with interest at 10%. The date at which the loan is made is considered time 0, and the time periods are measured in years from that date. The $10,000 amount of the loan is called the *principal* of the loan.

12.8.   Short-Term Notes

Plan 1 is typical of a method of repayment often used with renewable short-term notes. Suppose that you borrow some money on a 30-day note in order to meet your payroll. At the end of the month you could pay back the principal and the interest out of your draw, but then you would find that you had to borrow the same amount again. To save bookkeeping and certain other finance charges, the bank offers to let you continue the note, if you will pay only the interest due at the end of every year. This process is continued until the end of the job, at which time you agree to pay off the note plus the interest for the last month. Notice that in Plan 1 no money is paid on the principal until the very end of the total time period, but interest is paid at the end of each year.

TABLE 12.1. Four Plans for Repayment of $10,000 in 10 Years with Interest at 10%

End of Year	Interest Due (10% of Money Owed at Start of Year)	Total Money Owed Before Year-End Payment	Year-End Payment	Money Owed after Year-End Payment
Plan 1				
0				$10,000.00
1	$1,000.00	$11,000.00	$ 1,000.00	10,000.00
2	1,000.00	11,000.00	1,000.00	10,000.00
3	1,000.00	11,000.00	1,000.00	10,000.00
4	1,000.00	11,000.00	1,000.00	10,000.00
5	1,000.00	11,000.00	1,000.00	10,000.00
6	1,000.00	11,000.00	1,000.00	10,000.00
7	1,000.00	11,000.00	1,000.00	10,000.00
8	1,000.00	11,000.00	1,000.00	10,000.00
9	1,000.00	11,000.00	1,000.00	10,000.00
10	1,000.00	11,000.00	11,000.00	0
Plan 2				
0				$10,000.00
1	$1,000.00	$11,000.00	$2,000.00	9,000.00
2	900.00	9,900.00	1,900.00	8,000.00
3	800.00	8,800.00	1,800.00	7,000.00
4	700.00	7,700.00	1,700.00	6,000.00
5	600.00	6,600.00	1,600.00	5,000.00
6	500.00	5,500.00	1,500.00	4,000.00
7	400.00	4,400.00	1,400.00	3,000.00
8	300.00	3,300.00	1,300.00	2,000.00
9	200.00	2,200.00	1,200.00	1,000.00
10	100.00	1,100.00	1,100.00	0
Plan 3				
0				$10,000.00
1	$1,000.00	$11,000.00	$1,627.45	9,372.55
2	937.24	10,309.79	1,627.45	8,682.34
3	868.23	9,550.57	1,627.45	7,923.12
4	792.31	8,715.43	1,627.45	7,087.98
5	708.80	7,796.78	1,627.45	6,169.33
6	616.93	6,786.26	1,627.45	5,158.81
7	515.88	5,674.67	1,627.45	4,047.24
8	404.72	4,451.96	1,627.45	2,824.51
9	282.45	3,106.96	1,627.45	1,479.51
10	147.95	1,627.45	1,627.45	0
Plan 4				
0				$10,000.00
1	$1,000.00	$11,000.00	$ 0.00	11,000.00
2	1,100.00	12,100.00	0.00	12,100.00
3	1,210.00	13,310.00	0.00	13,310.00

TABLE 12.1. *(Continued)*

End of Year	Interest Due (10% of Money Owed at Start of Year)	Total Money Owed Before Year-End Payment	Year-End Payment	Money Owed after Year-End Payment
Plan 4				
4	1,331.00	14,641.00	0.00	14,641.00
5	1,464.10	16,105.10	0.00	16,105.10
6	1,610.51	17,715.61	0.00	17,715.61
7	1,771.56	19,487.17	0.00	19,487.17
8	1,948.72	21,435.89	0.00	21,435.89
9	2,143.59	23,579.48	0.00	23,579.48
10	2,357.95	25,937.43	25,937.43	0

12.9. Uniform Principal Payments

Plan 2 is another method often used in repaying short-term notes. It differs from Plan 1 in that periodic payments are made on the principal, in addition to paying the interest due at the end of each period of time. But in this case, because the principal amount is being steadily reduced, each of the interest payments is less than the preceding one. Thus the amount of dollars paid on each installment is a steadily decreasing number. In fact, the total number of dollars paid in Plan 1 is $20,000, but the total number of dollars paid in Plan 2 is $15,000.

12.10. Equal Periodic Payments

The method used in Plan 3 is the usual way of paying off a mortgage. The amount of $1627.45 is called the *equal periodic payment*. Out of this payment first comes all the interest due at that time, then the remainder is applied toward reducing the principal. For example, out of the first payment of $1627.45, the first $1000 is used to pay the interest, and the remaining ($1627.45 − $1000.00 =) $627.45 is paid on the principal, reducing it from $10,000 to $9372.55. From the second $1627.45 payment, the first $937.24 is paid on interest, leaving ($1627.45 − $937.24) = $690.21 to be paid on the principal. This reduces the principal balance from $9372.55 to $8682.34. Notice that at first more than half of the equal periodic payment is used to pay interest, but as time goes on more is applied to the principal and less to the interest, until at the end only $147.95 is due on interest, while $1479.50 is applied to the principal. But the number of dollars paid each time remains the same. The total number of dollars actually paid back is 10 × $1627.56 = $16,274.50, which includes $10,000 for repayment of principal and $6274.50 for the total of the interest payments.

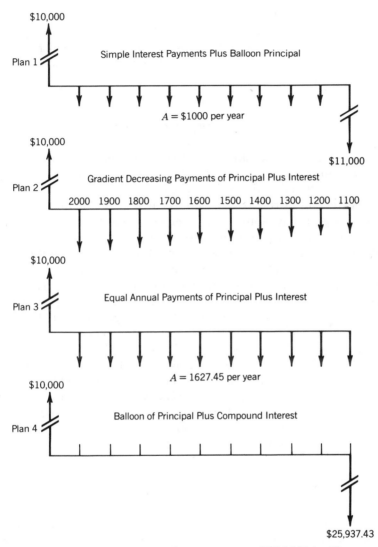

$10,000

Plan 1

Simple Interest Payments Plus Balloon Principal

$A = \$1000$ per year

$10,000

$11,000

Plan 2

Gradient Decreasing Payments of Principal Plus Interest

2000 1900 1800 1700 1600 1500 1400 1300 1200 1100

$10,000

Plan 3

Equal Annual Payments of Principal Plus Interest

$A = 1627.45$ per year

$10,000

Plan 4

Balloon of Principal Plus Compound Interest

$25,937.43

FIGURE 12.4. Cash flow diagram for repayment of $10,000 in 10 years with interest at 10%.

12.11. Paying Off a Loan

One last observation should be made concerning Plan 3. Unless a penalty clause is included in the terms of the loan such as that shown in the commitment described previously, the principal balance amount can be paid off at any time, thus discharging your debt. For example, after making the fifth payment of $1,627.45 there remains an unpaid balance of $6169.33.

At that time, if an additional payment of $6169.33 is made, the debt will be canceled. No further interest is owed. Interest is paid only on money that is in the borrower's possession for the time during which it is held. No interest is owed on money not received, nor is interest owed on money already paid back.

12.12.   Compound Interest

Plan 4 is an example of what is called *compound interest*. It is typical of what happens to a sum of money left untouched in a bank account. If considered as a debt, it is often called a *balloon note* because the lump sum "balloon" is there at the end of the loan period. (Plan 1 is another form of balloon note.) Because no payments whatsoever are made between the time the money is borrowed (or placed in the bank) and the end of a specified time period, interest is paid on the interest already accumulated in the account. For instance, at the end of the first year, because no payments are made out of the account, the total amount owed is the principal of $10,000 plus the interest of $1000 for a total of $11,000. At the beginning of the second year, then, $11,000 is owed and interest is charged against this amount during the course of the year. At the end of the second year the interest due is $1100 instead of $1000, the additional $100 being the interest on the interest ($0.10 \times \$1000 = \$100$). This means that interest is compounding. In the third year interest will be charged on $11,000 + $1100 = $12,100, and so on. At the end of 10 years the total compound amount of $25,937.43 must be paid to discharge the debt. If the debt is paid off sooner, say at the end of 5 years, the total amount due at that time must be paid ($16,105.10 for 5 years).

12.13.   Equivalence

Surprisingly, all of the four plans of repayment are equivalent, one to the other, with respect to the time period and the rate of interest. This is true because all four plans provide for payment of 10 cents of interest for every one dollar still owed at the end of each year. The dollar amounts are different because each plan has a different repayment schedule. For example, compare Plan 1 with Plan 4. At the end of the first year the lender bank receives a $1000 interest payment, which it can immediately lend to some third party (call him Jones) at the same rate of interest. During the second year that $1000 will earn $0.10 \times \$1000 = \$100$ for the bank. At the end of the second year the bank receives $1000 from the borrower using Plan 1, plus an additional $100 from Jones, for a total of $1100, the same as the interest shown in Plan 4. The bank, having lent $1000 to Jones, now lends $1100 to Smith, and receives as interest at the end of the third year

$1000 from the first borrower, $100 from Jones, and $0.10 \times \$1100 =$ $110.00 from Smith, for a total of $1000 + $100 + $110 = $1210.00, the same amount as shown in Plan 4. This process can be repeated all the way down the line for 10 years, so that Plan 1 is equivalent to Plan 4.

12.14. What Do All Four Plans Have in Common?

It is not difficult to see that the same line of reasoning can be applied when comparing Plan 2 with Plan 4, and that it will also hold true when considering Plan 3 in relation to Plan 4. In other words, each one of the first three plans is equivalent to Plan 4 from the lender's viewpoint, and thus they are all equivalent to each other provided that all funds earn interest at the same rate. The bank will make the same amount of money on the loan no matter which one of the four plans is adopted for repayment. There are three things common to each one of the plans: the principal of $10,000, the interest rate of 10%, and the total time of 10 years.

12.15. The Borrower's Viewpoint

From the borrower's viewpoint also, the plans are equivalent. Again consider Plan 1 versus Plan 4. The money you have to work with should make money for you, or you should not have borrowed it in the first place. In fact, it must earn at least as much as it costs to borrow. Therefore the borrowed $10,000 should earn at least $1000 for you during the first year. At the end of that year, if you do not pay $1000 to the bank, you should be able to put the money to work so that it earns at least $0.10 \times \$1000 = \$100$ for you during the second year. Simultaneously, the original $10,000 is earning at least $1000 for you during the second year, so that at the end of that year you should have a capital of $10,000 + $1000 + $100 + $1000 = $12,100, the same as the amount shown in Plan 4. It is simply a question of which pocket the money goes into. Therefore Plan 1 is equivalent to Plan 4 from the borrower's viewpoint, and it can be seen that, in a similar way, all the plans are equivalent to one another.

12.16. Superiority of Plans

There is no single answer as to which plan is superior. In practice the lender usually requires one specific method of repayment. Therefore if the borrower wants the money, he will have to agree to the terms. If the option is with the borrower, the decision depends on how he thinks he will best be able to pay the money back, and what use he has for the money. In these examples, the actual true cost to the borrower and the total income to the bank are the same, no matter which plan is adopted. The total number of dollars changing hands is not the same due to the timing of the repayments:

$20,000   for Plan 1

$15,000   for Plan 2

$16,274.50   for Plan 3

$25,937.43   for Plan 4

However, under each of the 4 plans for lending out $10,000, if all of the principal and interest payments received by the bank are immediately reinvested at 10%, then the bank would accumulate $25,937.43 by the end of year 10 in all four cases.

12.17.   Present Worth

All four plans discussed in the preceding paragraph were equivalent because they were based on the same loan amount (that is $10,000) at the beginning of the loan period (time zero) and yielded the same interest (10%) on the unpaid balance. Measured at the time the loan is made (or at the time the funds are invested) the amount of principal is called the Present Worth of the money and is designated as *PW,* or sometimes just *P* for short.

Example 12.4.   Find *P,* Given *F, i, n.*

A particular loan specifies a promise to repay a lump sum of $25,937 at a time 10 years from now, with 10% interest on the unpaid balance. Find the present worth of that future $25,937 lump sum payment.

Solution: $P = F(P/F, 10\%, 10)$

$\qquad\qquad = \$25,937 \times 0.38554$

$\qquad\qquad = \$10,000$

where $F = \$25,937$

and $(P/F, 10\%, 10) = 0.38554$ found either from solving equation (1b), or from finding the value in Appendix C, the 10% table, column headed *P/F,* line *n* = 10.

The present worth is $10,000. In other words $10,000 could be loaned at 10% interest compounded annually on this promise to repay $25,937 at EOY 10.

Example 12.5.   Find *P*, Given *A, i, n.*

Another loan specifies a promise to repay $1627.45 annually for each of the next 10 years, and the interest is 10% on the unpaid balance. Find the present worth of all ten of those repayments.

Solution: $P =$ $A(P/A, 10\%\ 10)$

$=$ $\$1627.45 \times 6.1445$

$=$ $\$10,000$

where $A =$ $\$1,627.45$ the annual payment

and $(P/A, 10\%, 10) =$ 6.1445 found either by solving equation (3b) or by finding the value in Appendix C, the 10% table, column headed *P/A, line n* = 10.

Again the present worth is $10,000. Or the amount that could be loaned with this promise to repay is $10,000 also. The method of calculating the sums involved is presented a few paragraphs further on.

12.18.   Meaning of "Present"

The word "present" in "present worth" refers to the date on which the loan is made, sometimes called problem-time-zero or PTZ for short. Thus, if money had been borrowed in the year 1426 with a promise to pay it back in 1492, then for purposes of computation 1426 becomes the "present" or PTZ, even though that date is more than 500 years ago. Or, if a loan is going to be made in the year 2437, to be paid back by 2447, the loan amount of PTZ in 2437 is its present worth. The rules of interest computation are concerned with accrued interest and are not to be confused with inflation or deflation.

12.19.   Future Value of Present Worth, F/P

The rules of compound interest lead to six useful equations, which will be designated as equations 1 through 6. The first of these can be derived fairly simply.

Suppose that a loan has a principal amount *P*. If the interest *rate* is *i*, then at the end of the first year the *amount* of interest owed will be *iP*. For instance, if the principal, $P = \$100$, and interest is at 10%, or $i = 0.10$, then the amount of interest owed is iP, or $0.10 \times \$100 = \$10$ interest at end of year 1, or at EOY 1. The *total* amount owed will be $P + iP = P(1 + i)$ or $\$100 \times 1.10 = \$110$ principal plus interest. If none of this is paid, at the end of the second year the amount of interest owed will be the rate of interest times the total amount owed at the beginning of the second year, which comes to $iP(1 + i)$. The total amount owed at the end of the second year will be the total owed at the beginning of the year plus the accrued interest for the year, which is

$$
\begin{aligned}
P(1 + i) + iP(1 + i) &= (P + iP)(1 + i) \\
&= P(1 + i)(1 + i) \\
&= P(1 + i)^2 \\
&= \$100 \times 1.10^2 \\
&= \$121 \text{ principal plus accrued interest compounded}
\end{aligned}
$$

At the end of the third year the interest owed will be $iP(1 + i)^2$. Add this to the amount owed at the beginning of the third year, and you get

$$
\begin{aligned}
P(1 + i)^2 + iP(1 + i)^2 &= (P + iP)(1 + i)^2 \\
&= P(1 + i)(1 + i)^2 \\
&= P(1 + i)^3 \\
&= \$100 \times 1.10^3 \\
&= \$133.10 \text{ principal plus accrued interest com-} \\
&\ \text{pounded}
\end{aligned}
$$

By extension, the rule now becomes obvious. If F is the total sum owed at the end of n years, then

$$
F = P(1 + i)^n \tag{1a}
$$

The expression $(1 + i)^n$ may be called the *single payment compound amount factor*, SPCAF, but more often is referred to as $(F/P, i, n)$, or just F/P, the ratio of future value F to present value P. (See Appendix C) For example, if the interest is 10% and $10,000 is borrowed for a period of 10 years, then

$$
\begin{aligned}
F &= \$10,000\,(1 + 0.10)^{10} \\
&= \$10,000 \times 2.593743 \\
&= \$25,937.43
\end{aligned}
$$

will be the total amount owed at the end of 10 years, if no payments are made before then (see Plan 4, Table 12.1). Notice that the interest rate must be put into decimal form to use in the formula. By dividing both sides of equation 1 by $(1 + i)^n$, P in terms of F is obtained.

$$P = \frac{F}{(1 + i)^n} \qquad (1b)$$

The term $1/(1 + i)^n$ is sometimes called the *single payment present worth factor*, SPPWF, but more often is referred to as *(P/F, i, n)*, or just *P/F*, where *P/F* is the ratio of present value P to future value F for any given values of i and n. For convenience of notation, this factor may be designated by the notation *(F/P, i, n)*. Using this notation equations 1a and 1b would be expressed as

$$F = P(F/P, i, n) \qquad (1a)$$

$$P = F(P/F, i, n) \qquad (1b)$$

Example 12.6.   Find P Given F, i, n.

If 10 years from now a proposed investment will repay you a lump sum amount of \$25,937.43, how much can you afford to invest now (Present Worth) if an acceptable interest rate is 10%? In other words, how much would you have to invest in an account drawing 10% interest compounded annually in order to accumulate \$25,937.43 in 10 years?

Solution: Using equation (1b), the following solution is obtained.

$$P = \frac{\$25,937.43}{(1 + 0.10)^{10}}$$

$$= \frac{\$25,937.43}{2.593743}$$

$$= \$10,000.00$$

Or using the 10% table and the *P/F* column from Appendix C,

$$P = F(P/F, 10\%, 10)$$

$$= \$25,937.43 \times 0.38554$$

$$= \$10,000$$

12.20. Equal Periodic Payments, A Derivation of the F/A Equation

Suppose that you want to have F dollars in the bank at the end of n years, and you want to make a deposit of A dollars at the end of each year such that the deposits will compound to F. No deposit is made at the beginning of the time period, at time 0, and the first deposit is made at the end of the first year (EOY 1). Then that first deposit will accumulate interest, not for n years, but for $(n - 1)$ years; since it earns no interest during the first year, because it was deposited, not at time 0, but rather at the end of the first year. At the end of n years, if the interest rate is i, the first deposit will amount to

$$A(1 + i)^{n-1}$$

The second deposit is of the same amount of A dollars, but it is made at the end of the second year (EOY 2), so that it will amount to

$$A(1 + i)^{n-2}$$

The third year's deposit will accumulate to

$$A(1 + i)^{n-3}$$

and so on. The deposit just before the last one will be

$$A(1 + i)$$

and the last deposit will earn no interest—it will simply be equal to A dollars.
    The total in the bank will be the sum of all the future sums:

$$F = A(1 + i)^{n-1} + A(1 + i)^{n-2} + A(1 + i)^{n-3} + \cdots + A(1 + i) + A$$

Reversing the order and factoring gives

$$F = A[1 + (1 + i) + (1 + i)^2 + \cdots + (1 + i)^{n-3} \\ + (1 + i)^{n-2} + (1 + i)^{n-1}]$$

Now multiply both sides of the equation by $(1 + i)$ to get

$$F + iF = A[(1 + i) + (1 + i)^2 + \cdots + (1 + i)^{n-3} \\ + (1 + i)^{n-2} + (1 + i)^{n-1} + (1 + i)^{n}]$$

Subtract the first equation from the second, and the only terms remaining are

$$iF = A[-1 + (1 + i)^n]$$

or

$$iF = A[(1 + i)^n - 1]$$

and

$$F = A\frac{(1 + i)^n - 1}{i} \qquad (2a)$$

The expression $F/A = \dfrac{(1 + i)^n - 1}{i}$ is called the "uniform annual series compound amount factor," UACAF, or just F/A, or $(F/A, i, n)$.
Then

$$A = F\left[\frac{i}{(1 + i)^n - 1}\right] \qquad (2b)$$

The quantity $i/[(1 + i)^n - 1]$ is sometimes called the *sinking fund deposit factor*, SFF, or just A/F. For convenience of notation this factor may be designated by the notation $(A/F, i, n)$. Using this notation, equation 2 is expressed as:

$$F = A(F/A, i, n) \qquad \text{or} \qquad A = F(A/F, i, n)$$

Example 12.7   Find A, Given F, i, n.

Determine how much must be deposited at the end of each year for 10 years in order to accumulate \$25,937.43 if the interest is 10%.

Solution: $A = F\left[\dfrac{i}{(1 + i)^n - 1}\right]$

$$= \$25,937.43\left[\frac{0.10}{(1 + 0.10)^{10} - 1}\right]$$

$$= \$25,937.43 \times \frac{0.10}{(2.158924) - 1}$$

$$= \$25,937.43 \times 0.06274539$$

$$= \$1627.45$$

or alternatively using Appendix C, the 10% table and the *A/F* column on the line $n = 10$:

$$A = F(A/F \ 10\% \ 10)$$

$$= \$25,937.43 \times 0.06275$$

$$= \$1627.57$$

The difference of $0.12 is due to rounding error of course. (See Plan 3, Table 12.1 and note that it is indeed equivalent to Plan 4.)

12.21. Capital Recovery Factor, A/P

Because $F = P(1 + i)^n$, it is possible to substitute in equation 2 and get

$$A = P(1 + i)^n \left[\frac{i}{(1 + i)^n - 1} \right]$$

or

$$A = P \left[\frac{i(1 + i)^n}{(1 + i)^n - 1} \right] \tag{3a}$$

Another form of equation 3 which gives exactly the same answer is

$$A = P \left[\frac{i}{(1 + i)^n - 1} + i \right]$$

For convenience of notation equation 3 may be expressed as

$$A = P(A/P, i, n) \quad \text{and} \quad P = A(P/A, i, n)$$

Equation 3 answers the question: How much must be paid at the end of each year so as to pay off a debt of *P* dollars in *n* years if interest is at rate *i*? Alternatively, the question can be asked as follows: How much must I get at the end of each year for *n* years to justify an investment of *P* dollars if I want an *i* rate of return on my investment? As used with mortgages, the expression

$$A/P = \frac{i(1 + i)^n}{(1 + i)^n - 1}$$

is referred to as the *capital recovery factor* or just *A/P*. Because both forms of equation 3 produce exactly the same answer, the capital recovery factor is always equal to the sinking fund factor plus the interest rate:

$$\frac{i(1 + i)^n}{(1 + i)^n - 1} = \frac{i}{(1 - i)^n - 1} + i$$

or

$$A/P = A/F + i$$

Example 12.8.   Find *A*, Given *P*, *i*, *n*.

What are the annual payments on a 10-year mortgage of $10,000 if the interest is 10%?

Solution: The yearly payment would be

$$A = P\left[\frac{i(1 + i)^n}{(1 + i)^n - 1}\right]$$

$$= \$10,000 \left[\frac{0.10 \, (1 + 0.10)^{10}}{(1 + 0.10)^{10} - 1}\right]$$

$$= \$10,000 \left(\frac{0.10 \times 2.593743}{2.593743 - 1}\right)$$

$$= \$10,000 \times 0.162745$$

$$= \$1,627.45$$

or, alternatively, using the 10% table and the *A/P* column:

$$A = P(A/P, 10\%, 10)$$

$$= \$10,000 \times 0.16275$$

$$= \$1,627.50$$

If monthly payments are specified, then the 10% interest becomes a nominal annual interest rate, and the actual *i* is found as

$$i = \frac{0.10}{12} = 0.008333/\text{month}$$

Equation 3a can be inverted to obtain

$$P = A\left[\frac{(1 + i)^n - 1}{i(1 + i)^n}\right] \qquad (3b)$$

designated as $P = A(P/A, i, n)$

The expression $P/A = \dfrac{(1 + i)^n - 1}{i(1 + i)^n}$ is the *uniform annual series present worth factor*, UAPWF, or just P/A or $(P/A, i, n)$.

Example 12.9.   Find P, Given A, i, n.

Find how large a loan can be obtained upon a promise to pay $1627.45 per year for 10 years if the interest is 10%.

Solution: Using equation (3b) we have:

$$P = A\left[\frac{(1 + i)^n - 1}{i(1 + i)^n}\right]$$

$$= \$1627.45\left[\frac{(1 + 0.10)^{10} - 1}{0.10\,(1 + 0.10)^{10}}\right]$$

$$= \$1627.45\left(\frac{2.593743 - 1}{0.10 \times 2.593743}\right)$$

$$= \$1627.45 \times 6.144568$$

$$= \$10,000$$

or, alternatively, turning to Appendix C, the 10% tables, and the P/A column and line $n = 10$:

$$P = A(P/A, 10\%, 10)$$

$$= \$1627.45 \times 6.1446$$

$$= \$10,000$$

To facilitate the solution of problems, tables of all six factors are given in Appendix C.

12.22. Arithmetic Gradient, G

Periodic payments that increase or decrease are encountered frequently. For instance operating and maintenance costs for equipment or structures may increase every year, productivity of equipment may decrease with age, rental incomes or expenditures may increase every year. These increases or decreases may be reasonably predictable and may be approximated by an arithmetic gradient, G.

For example, annual maintenance costs on a proposed mechanical unit are estimated at

Year	Maintenance
1	0
2	1000
3	2000

The growth increment in this example is G = $1000/year and is termed a "gradient." Since the increment is a *constant amount* at $1000, the progression is described as arithmetic (compared to geometric if the growth increased by the same *percent* each year).

This situation occurs often enough to warrant the use of special equivalence factors relating the arithmetic gradient, G, to other cash flows. The gradient G can be related to the future value, F, the uniform series value, A, and the present value, P.

12.23. Future Values of Gradient Amounts, F/G

Notice that the maintenance table above starts with a value of $0 at year 1 and then increased by a gradient G of $1000/year. The cash flow for such a series is shown in Figure 12.5.

The gradient amount, designated as G in the generalized problem, can be either positive or negative, but always increases in absolute value with increasing time (the values grow larger from left to right on the cash flow diagram) as shown in Figure 12.5.

The equations for F/G, A/G, and P/G are derived in a manner similar to the time value equations derived previously, resulting in the relationships shown in Figure 12.6.

The use of these equations are shown in the following examples.

Example 12.10. Find F, Given G, i, and n.

A contractor is putting money aside to purchase some equipment, deciding to invest $2000 at EOY 2, and increasing the investment amount by an

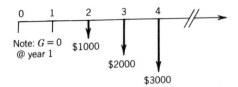

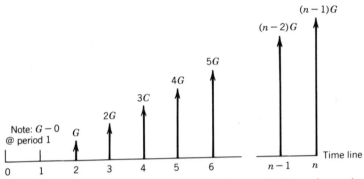

FIGURE 12.5.   Cash flow diagram for $G = -\$1000$ and $r = 4$.

additional $2000 each year until EOY 5, as shown in Figure 12.7. The investment will earn 10% compounded annually. Find the balance in the account after 5 years.

Solution: The problem may be solved by either of the following methods:
a.   Solution by equation.

$$F = \frac{G}{i}\left[\frac{1 + i)^{n} - 1}{i} - n\right]$$

$$= \frac{2000}{0.10}\left[\frac{(1 + 0.10)^{5} - 1}{0.10} - 5\right]$$

$$= \$22{,}102$$

b.   Solution by factor from the tables.

$$F = G(F/G,\ 10\%,\ 5)$$

$$= \$2000 \times 11.051$$

$$= \$22{,}102$$

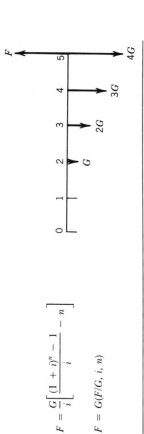

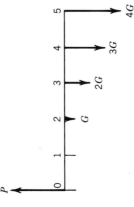

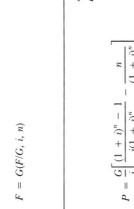

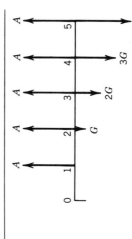

Future worth, F, of an
arithmetic gradient series, G
(n = Number of payments plus 1)
Notation form

$$F = \frac{G}{i}\left[\frac{(1 + i)^n - 1}{i} - n\right]$$

$$F = G(F/G, i, n)$$

Present worth, P, of an
arithmetic gradient series, G
(n = Number of payments plus 1)
Notation form

$$P = \frac{G}{i}\left[\frac{(1 + i)^n - 1}{i(1 + i)^n} - \frac{n}{(1 + i)^n}\right]$$

$$P = G(F/G, i, n)$$

Periodic uniform series, A,
equivalent of arithmetic
gradient series, G
(n = Number of payments plus 1)
Notation form

$$A = G\left[\frac{1}{i} - \frac{i}{(1 + i)^n - 1}\right]$$

$$A = G(A/G, i, n)$$

FIGURE 12.6. Cash flow diagrams illustrating typical arithmetic gradients (G).

185

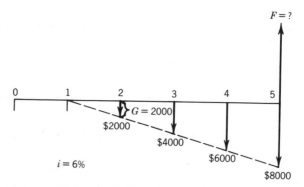

FIGURE 12.7.   Cash flow diagram for example 12.10.

12.24. Cash Flow Diagram and Graph of Increase in F as n Increases

The graph in Figure 12.8 illustrates the timing of the cash flows of G, as well as the relationships between G, F, i, and n, as illustrated by Example 12.10. Note that the first deposit of G ($2000 in this example) occurs at the *end* of the *second* period. Therefore the first interest is not earned until the end of the third period. Gradient deposits are made at the end of every year except the first year, so there are a total of $(n-1)$ deposits in the G

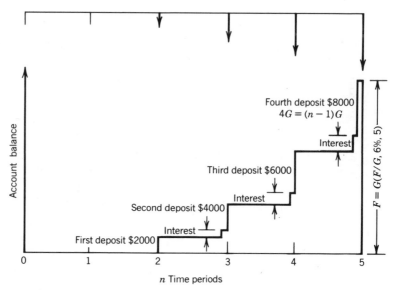

FIGURE 12.8.   Balance in the account, corresponding to the cash flow diagram for example 12.10.

series (as contrasted to n deposits in all of the A series). Deposits are increased by the amount G ($2000 in this case) for each period (year for this example). Therefore, the amount of the last deposit is $G(n-1)$, or in this example $2000 (5-1) = 8000. An ability to determine the correct value of n is essential in solving many problems commonly encountered. Since there are always $(n-1)$ gradient deposits in the G series, the value of n may be determined as the number of deposits plus one.

Example 12.11.   Find Decreasing Gradient G and A, in Terms of Dollars per Cubic Yard, Given F, i, and, n, with G and A in Terms of Cubic Yards per Year.

A contractor estimates that 10 years from today he will need a total of $80,000 in order to replace a dragline he now owns. He expects to move about 100,000 cubic yards of earth this year with the dragline, but due to age and increasing downtime he expects this amount to decrease by about 5000 cubic yards per year each year over the 10-year period. His other operating and maintenance (O & M) costs run about $0.50 per cubic yard. He needs to determine the additional charge per cubic yard required to accumulate the $80,000 in 10 years if he can invest the amount earned from the additional charge at the end of each year and earn 10% interest compounded.

Solution: This problem is approached by separating it into simple components, solving each component, and summing the results. First identify the A series and separate it from the G series. Then solve each series independently and add up the total. The variables are identified as:

$F = \$80,000$

$n = 10$ years

$A = (100,000$ cubic yards per year$) \times ($dollars per cubic yard$)$

$G = (-5000$ cubic yards per year$) \times ($dollars per cubic yard$)$

$i = 10\%$

The "other O & M costs" of $0.50/cubic yard is extraneous information not needed to solve the problem. The inclusion of such information here simulates the real world where myriads of useless information is often available to confuse the problem.

One basic problem here is translating cubic yards/year into dollars/year. This may be done by letting y represent the extra charge in terms of dollars per cubic yard. Then find how many dollars/cubic yard $\times$ cubic yards/year is needed to accumulate the $80,000.

A = (100,000 cubic yards) × (y), the annual base income

G = (-5000 cubic yards) × (y) the declining gradient

$F = A(F/A, i, n) - G(F/G, i, n)$

$$\$80,000 = (100,000 \, y)(F/A, 10\%, 10) - \frac{(5000 \, y)}{0.10}\left[\frac{(1.10)^{10} - 1}{0.10} - 10\right]$$

$$\$80,000 = 1,593,740 \, y - 296,871 \, y$$

$$y = \frac{80,000}{1,296,869}$$

$$= \$0.0617$$

$$= \$0.062/\text{cubic yard}$$

By charging an extra 6.2 cents per cubic yard and investing the resulting funds at the end of each year at 10% compounded, the contractor will accumulate $80,000 at the end of 10 years to purchase a new dragline.

12.25.   Uniform Series Values of Gradient Amounts, A/G

Many times the problem involves finding an equivalent uniform series A that is equivalent to a gradient series G. The cash flow diagram for this problem is given in Figure 12.9.

The example that follows illustrates the A/G relationship.

Example 12.12.   Find G, Given P, A, and n.

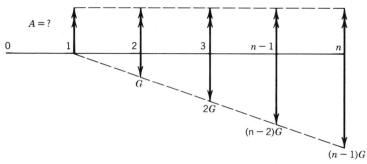

FIGURE 12.9.   A typical A/G equivalency.

Construction of an apartment project is estimated to cost $40,000 per apartment unit. Financing is available at 11% for 25 years on the full $40,000 cost. Due to competition in the area, rents must begin at $300 per month but may be raised some every year. Operating costs, taxes, and insurance amount to $100 per month, leaving $200 per month to meet principal and interest on the mortgage payments at the start. What annual increase in rents is necessary to pay off the 11%, 25-year mortgage, assuming the lender is willing to take arithmetic gradient payments with 11% interest on the unpaid balance. To simplify the calculations, use an annual basis instead of monthly.

Solution: The annual mortgage payments to amortize the $40,000 cost are found as:

$$A_1 = \$40,000 \ (A/P, \ 11\%, \ 25) = \$4748 \text{ per year}$$
$$0.1187$$

The initial rental income available to pay the mortgage is:

$$A_2 = \$200 \times 12 \text{ months} = \$2400 \text{ per year}$$

The equivalent annual payment deficit to be made up by periodic rent increase is:

$$\text{Deficit} = A_1 - A_2 = \$4748 - 2400 = \$2348 \text{ per year}$$

The required annual gradient increase in rent is found as:

$$G = \$2348 \ (G/A, \ 11\%, \ 25) = \$330.50 \text{ per year}$$

(Equivalent to raising the rent approximately $330.50 per year = $27.54 per month each year from a base of $300 per month the first year.)

12.26.   *Present Worth Values of Gradient Amounts (P/G)*

This *P/G* series provides for solution of problems involving a payment (cost or income) that changes (increase or decreases) by *G* dollars per period for *n* periods. Interest is earned on any balance that remains in the loan or deposit. The equation is derived by finding the present sum required to finance this *G* series for *n* periods and results in a balance-in-the-account diagram as shown in Figure 12.10.

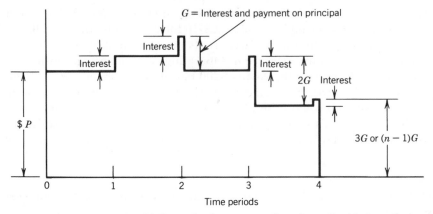

FIGURE 12.10. Graph of balance in the account for a loan P with $(n - 1)$ repayments periodically increasing by amount G.

Example 12.13. Find P, Given G, i, and n.

The repair costs on a water pump are expected to be zero the first year and to increase \$100 per year over the 5-year life of the pump. A fund bearing 6% on the remaining balance is set up in advance to pay for these costs. How much should be in the fund? Costs are billed to the fund at the end of each year.

Solution: The cash flow diagram appears in Figure 12.11.

a. Solution by equation:

$$P = \frac{G}{i}\left[\frac{(1 + i)^n - 1}{i(1 + i)^n} - \frac{n}{(1 + i)^n}\right]$$

$$= \frac{100}{0.06}\left[\frac{(1 + 0.06)^5 - 1}{0.06\,(1 + 0.06)^5} - \frac{5}{(1 + 0.06)^5}\right]$$

$$= 100 \times 7.9345$$

$$= \$793.45$$

b. Alternative solution by use of the tables:

$$P = G\,(P/G,\,6\%,\,5)$$

$$= 100\,(7.9345)$$

$$= 793.45$$

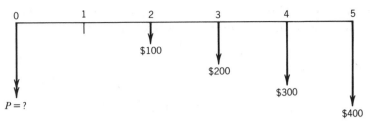

FIGURE 12.11. Cash flow diagram for example 12.13.

12.27. Periods Shorter Than One Year

All six equations (1 through 6) assume that n is the number of compounding periods, whether months, quarters, years or whatever, while i is the interest rate at which interest is compounded on the unpaid balance. To use the equations for any compounding periods, whether longer or shorter than a year, simply make the following substitutions:

n = number of *compounding periods*

i = interest rate *per compounding period*

A = amount paid at end of *each period*

F = sum of money at end of the total *number of periods*

P remains the present worth, to be measured at the beginning of the reckoning.

Example 12.14. Find how much must be be paid back at the end of 1 year, if $10,000 is borrowed at the beginning of that year and interest is 10% compounded *semiannually* (every 6 months).

Solution: Here $n = 2$; therefore $i = 0.10/2 = 0.05$, and

$$F = P(1 + i)^n$$
$$= \$10,000 \, (1 + 0.05)^2$$
$$= \$10,000 \times 1.1025$$
$$= \$11,025$$

or

$$F = P(F/P, \, 5\%, \, 2)$$
$$= \$10,000 \, (1.1025)$$
$$= \$11,025$$

Example 12.15.   Find how much must be paid back at the end of 10 years given the same terms and conditions as Example 12.14.

Solution: We have $n = 10 \times 2 = 20$, $i = 0.10/2 = 0.05$, and

$$F = \$10,000 \, (1 + 0.05)^{20}$$
$$= \$10,000 \times 2.653298$$
$$= \$26,532.98$$

or

$$F = P(F/P, 5\%, 20)$$
$$= \$10,000 \, (2.6533)$$
$$= \$26,533$$

This is \$595.55 more than the \$25,937.43 determined by compounding at 10% for 10 years, since the compounding occurs twice as often but for only one-half as much interest rate each time.

12.28.   *Effective Rate*

The effective rate of interest can be found by using the equation

$$i_e = (1 + i)^n - 1$$

where i and n are the numbers obtained from the procedure used in Example 12.14 and 12.15. For instance, the effective rate of interest for 10% compounded semiannually becomes

$$i_e = (1 + 0.05)^2 - 1 = 1.1025 - 1 = 0.1025$$

or 10.25%. Similarly, the true annual rate corresponding to 1% per month would be

$$i_e = (1 + 0.01)^{12} - 1 = 1.1268 - 1 = 0.1268 \text{ or } 12.68\%, \text{ not } 12\%.$$

There are no tables for these odd rates of interest, so the calculations must be run by solving the equation, or by interpolation of the tables.

12.30. *Examples of Typical Applications*

Example 12.16.   Find F, Given P, i and n.

Three typical examples of situations involving this approach are the following: (1) If $3000 is invested now at 5%, how much will it accumulate to in 20 years? or (2) What is the compound amount of $3000 for 20 years with interest at 5%? or (3) How much must be saved 20 years from now in order to justify a present expenditure of $3000 if money is worth 5%?

Solution: The cash flow diagram is shown in Figure 12.12.

$$i = 0.05$$
$$n = 20$$
$$P = \$3000$$
$$F = ?$$
$$F = P(1 + i)^n$$
$$= \$3000 \, (1 + 0.05)^{20}$$

The compound amount factor $(1.05)^{20}$ is given in the 5% table (Appendix C) as 2.653, so that

$$F = \$3000 \times 2.653$$
$$= \$7950$$

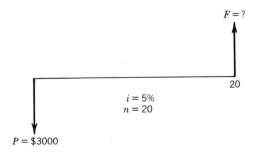

FIGURE 12.12.   Cash flow diagram for example 12.16.

or using the table for $i = 5\%$, the F/P column, and the row where $n = 20$, then

$$F = (F/P, 5\%, 20)$$
$$= \$3000 \times 2.6533$$
$$= \$7950$$

Example 12.17.   Find F, Given Several P, i, and n Values.

If $3000 is invested now, $2000 is invested 3 years from now, and $1000 is invested 6 years from now, all at 6%, what will be the total amount 15 years from now?

Solution: The cash flow diagram is drawn as shown in Figure 12.13.
   The formulas involving A cannot be used because the amounts are not all the same, the time intervals are not 1 year or less, and the first investment was made at the beginning of the time instead of at the end of the first time interval. The solution therefore requires three separate calculations involving P and F. The first "present" is now; the second "present" starts 3 years from now; and the third "present" begins 6 years from now.

$$P_1 = \$3000, P_2 = \$2000, P_3 = \$1000$$

$$n_1 = 15, n_2 = 12, n_3 = 9$$

$$i_1 = 0.06, i_2 = 0.06, i_3 = 0.06$$

$$F = F_1 + F_2 + F_3$$
$$= P_1(1 + i)^{n_1} + P_2(1 + i)^{n_2} + P_3(1 + i)^{n_3}$$
$$= \$3000(1.06)^{15} + \$2000(1.06)^{12} + \$1000(1.06)^9$$
$$= \$3000 \times 2.379 + \$2000 \times 2.012 + \$1000 \times 1.689$$
$$= \$7191 + \$4024 + \$1689$$
$$= \$12,904$$

Example 12.18.   Find F, Given P, i, and n Semiannual Compounding.

If $500 is deposited in an account now and the account bears interest at 12% (nominal) compounded semiannually, find the balance of accrued interest plus principal at the end of 10 years.

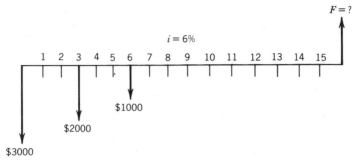

FIGURE 12.13. Cash flow diagram for example 12.17.

Solution: The cash flow diagram is shown in Figure 12.14.

The interest rate period is $12\%/2 = 6\%$. The number of periods is $10 \times 2 = 20$. From the 6% table (Appendix C) the compound amount factor is found to be 3.2071.

$$i = 0.06$$

$$n = 20$$

$$P = \$500$$

$$F = ?$$

$$F = P(F/P\ 6\%\ 20)$$

$$= \$500 \times 3.2071$$

$$= \$1,603.60$$

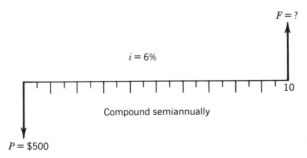

FIGURE 12.14. Cash flow diagram for example 12.18.

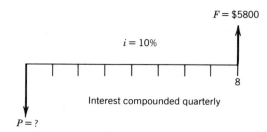

$F = \$5800$

$i = 10\%$

8

Interest compounded quarterly

$P = ?$

FIGURE 12.15.   Case flow diagram for example 12.19.

Example 12.19.   Find P, Given F, i, and n Compounded Quarterly.

What is the present worth of $5800 due 8 years from now if interest is 10% compounded quarterly?

Solution: The cash flow diagram is drawn in Figure 12.15.

$$i = 0.025$$

$$n = 32$$

$$F = \$5800$$

$$P = ?$$

$$P = F\left[\frac{1}{(1 + i)^n}\right]$$

$$= \$5800 \left[\frac{1}{(1.025)^{32}}\right]$$

$$= \$5800 \times 0.4538$$

$$= \$2632.04 \text{ or } \$2632$$

Example 12.20.   Find n, Given F/P and i.

How long will it take an investment of $1000 to increase to $2000 if the interest is 6%? In other words, how many years will it take for money to double itself with the interest at 6%?

Solution: The cash flow diagram is shown in Figure 12.16.

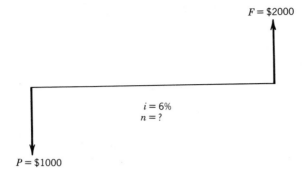

FIGURE 12.16.   Cash flow diagram for example 12.20.

$$i = 0.06$$

$$\frac{F}{P} = \frac{\$2000}{\$1000} = 2$$

$$n = ?$$

$$F = P(1 + i)^n$$

$$\frac{F}{P} = (1 + i)^n$$

$$2 = (1.06)^n$$

From the 6% column (Appendix C) the compound amount factors that bracket the derived value of $F/P = 2$ are:

F/P	n
1.898	11
2.000	?
2.012	12

The answer can be found by interpolation as

$$n = 11 \text{ years} + \frac{2.000 - 1.898}{2.012 - 1.898} \times 1 \text{ year}$$

$$n = 11.89 \text{ years}$$

or, many pocket calculators will solve directly for

$$n = \frac{\ln 2}{\ln 1.06} = 11.90 \text{ years}$$

12.31.   Rule of 70's

An approximate solution can also be found for the preceding problem by applying the rule of 70. The number 70 divided by the interest rate in percent will give the approximate number of years required for capital to double at that rate. If the interest is 6%,

$$\frac{70}{6} = 11.7 \text{ years (approximate)}$$

If the interest is 3%

$$\frac{0.70}{0.03} = \frac{70}{3} = 23.3 \text{ years, and so forth}$$

Conversely, if the number of years is given over which any investment has doubled in value, the interest rate may be approximated.

Example 12.21.   Find *F*, Given *P*, *i*, *n* Using the Approximate Rule of 70's.

   Land was purchased 5 years ago for $10,000 and is now worth $20,000. Find the compounded interest rate at which the value of the land increased each year.

Solution:

$$\frac{70}{5 \text{ years}} = 14\% \text{ compounded}$$

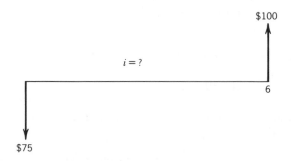

FIGURE 12.17.   Cash flow diagram for Example 12.22.

12.32. More Example Problems

Example 12.22.   Find i, Given F, P, and n.

A savings certificate that costs \$75 now will pay \$100 in 6 years. What is the interest rate?

Solution: The cash flow diagram is drawn in Figure 12.17.

$$\frac{F}{P} = \frac{100}{75} = 1.333$$

$$n = 6$$

$$i = ?$$

$$F/P = (1 + i)^n$$

$$\text{Since } F/P = \frac{100}{75} = 1.333$$

$$\text{Then } 1.333 = (1 + i)^6$$

$$1 + i = 1.333^{1/6}$$

$$i = 1.333^{1/6} - 1$$

$$i = 0.0491$$

$$i = 4.91\%$$

Example 12.23.   Find A, Given F, i, and n.

Three situations to which these conditions would apply are the following: (1) It is desired to establish a sinking fund that will accumulate to \$600,000 in 25 years. If interest is at 12%, how much must be deposited in equal annual payments at the end of each of those 25 years? or (2) What uniform annual expenditure, such as on preventive maintenance, is justifiable for each of 25 years in order to avoid spending \$600,000 at the end of that time, if money is worth 12%? or (3) What annual deposits must be made at 12% to save up enough to replace a structure that will cost \$600,000 just 25 years from now?

Solution: The cash flow diagram is shown in Figure 12.18.

$$i = 0.12$$

$$n = 25$$

$$F = \$600,000$$

$$A = ?$$

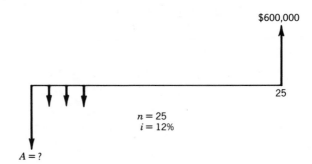

FIGURE 12.18.   Cash flow diagram for example 12.23.

The 12% table (Appendix C) gives the A/F, factor as 0.00750 for $n = 25$.

$$A = F(A/F, 5\%, 25)$$
$$= \$600,000 \, (0.00750)$$
$$= \$4500 \text{ per year}$$

Example 12.24.   Find F, Given A, i, and n.

How much would be accumulated in the sinking fund of Example 12.22 at the end of 15 years?

Solution: The cash flow diagram is drawn as shown in Figure 12.19.

$$i = 0.12$$
$$n = 15$$
$$A = \$4500/\text{year}$$
$$F = ?$$

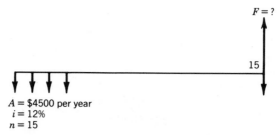

FIGURE 12.19.   Cash flow diagram for example 12.24.

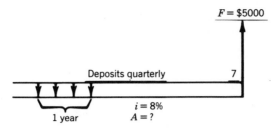

FIGURE 12.20. Cash flow diagram for example 12.25.

The 12% table gives the uniform series F/A factor for $n = 15$ as 37.279:

$$F = A(F/A, 5\%, 15)$$
$$= \$4500\ (37.279)$$
$$= \$167,755$$

Example 12.25. Find A, Given F, i, n with Quarterly Compounding.

What quarterly deposit must be made into a sinking fund to amount to $5000 in 7 years if interest is at 8% compounded quarterly?

Solution: The cash flow diagram is drawn in Figure 12.20.

$$i = 1/4 \times 0.08 = 0.02$$
$$n = 4 \times 7 = 28$$
$$F = \$5000$$
$$A = ?$$

From the 2% table (Appendix C) the A/E factor for $n = 28$ is found to be 0.02699.

$$A = F\ (A/F, 2\%, 28)$$
$$= \$5000\ (0.02699)$$
$$= \$134.95$$

Notice that this is a quarterly, not a yearly, deposit.

Example 12.26. Find A, Given P, i, and n.

What equal annual year-end payment of principal plus interest for 10 years is necessary to repay a loan of $10,000 if interest is 8%?

or

If $10,000 is deposited now at 8% interest, what uniform amount could be withdrawn at the end of each year for 10 years to have nothing left in the account at the beginning of the eleventh year?

Solution:

$$P = \$10,000$$

$$i = 0.08$$

$$n = 10 \text{ years}$$

$$\textit{Find } A = ?$$

$$A = P \ (A/P, \ 8\%, \ 10)$$

$$= \$10,000 \ (0.1490)$$

$$= \$1490.30 \text{ per year}$$

Example 12.27. Find the Balance Due on an Installment Loan. Find P, Given A, i, and n.

How much will be owed on the loan of Problem 12.26 after three payments have been made?

 As previously explained, only the remaining principal of the loan is owed at that point. No further interest is owed if the loan is paid off in one lump sum at this time. Therefore it is necessary to find the present worth of the loan at that time, which equals the present worth of the remaining payments.

Solution: The solution to Example 12.27 is the answer to this question: "What is the present worth of $1490.30 for 7 years with interest at 8%?"

 The cash flow diagram is drawn as shown in Figure 12.21.

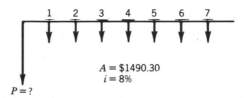

FIGURE 12.21.   Cash flow diagram for example 12.27.

$$i = 0.08$$

$$n = 7$$

$$A = \$1490.30$$

$$P = ?$$

$$P = A \left[\frac{(1 + i)^n - 1}{i(1 + i)^n} \right]$$

From the 8% table (Appendix C), the uniform annual series P/A factor is seen to be 5.206, so that

$$P = A(P/A, 9\%, 7)$$

$$= \$1490.30 \ (5.206)$$

$$= \$7758.50$$

Example 12.28.   Find the Cost of Financing a Loan. Find P, Given A, i, and n.

A construction manager has completed a certain job on which his fee was to have been $100,000. The owner offers to pay him $30,000 at once, and the remaining $70,000 in five yearly installments of $14,000 each. If the construction manager has to pay 12% interest on any money he borrows, how much is he losing by accepting the owner's offer?

Solution: The cash flow diagram is drawn in Figure 12.22.
The loss must be computed on the basis of what is owed now, not on what might happen in the future. True, the construction manager will have to pay 12% interest if the money must be borrowed, but if the money had been on hand, it could have been invested so as to earn 12%. Therefore

$$i = 0.12$$

$$n = 5$$

$$A = \$14,000$$

$$P = ?$$

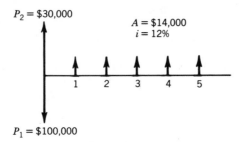

FIGURE 12.22. Cash flow diagram for example 12.28.

From the 12% table (Appendix C), the uniform annual series P/A factor is found to be 3.60478

$$P = A(P/A, 12\%, 5)$$
$$= \$14,000 \ (3.60478)$$
$$= \$50,467$$

Therefore the loss would be $\$70,000 - \$50,467 = \$19,533$, in terms of present worth dollars.

Example 12.29. Find A, Given P, i, and n.

What would be the proper payments for the owner to make under the conditions of Example 12.8?

Solution: The cash flow diagram is drawn as shown in Figure 12.23.

$$i = 0.12$$
$$n = 5$$
$$P = \$70,000$$
$$A = ?$$

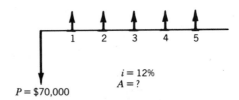

FIGURE 12.23. Cash flow diagram for example 12.29.

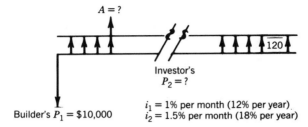

FIGURE 12.24.   Cash flow diagram for example 12.30.

From the 12% table, the *A/P* factor is found to be 0.27741, so the payments should be as follows:

$$A = P(A/P, \ 12\%, \ 5)$$
$$= \$70,000 \ (0.27741)$$
$$= \$19,419$$

instead of the $14,000 offered.

Example 12.30.   Find the Total *P*, Given *A*, *F*, and *n*.

A certain builder takes back a second mortgage for $10,000 amortized at 12% over a 10-year period. That is, the mortgage holder receives sufficient monthly payments to pay out the mortgage over a 120-month (10 years) period with interest on the unpaid balance at 1% per month compounded (12% nominal). After 4 months the builder wants to sell the second mortgage, but a prospective investor–buyer wants to receive an 18% return on his money. How much will the investor–buyer pay for the mortgage?

Solution: The cash flow diagram is drawn as shown in Figure 12.24. The builder's second mortgage calls for monthly payments of

$$A = P_1 \ (A/P, \ i, \ n)$$
$$= \$10,000 \ (A/P, \ 1\%, \ 120)$$
$$= \$10,000 \ (0.01434)$$
$$= \$143.47 \ \text{per month}$$

The investor is purchasing a mortgage that is 4 months old and only has 116 payments left. Furthermore the investor insists on $1\frac{1}{2}\%$/month (18% nominal), so to the investor the mortgage is worth:

$$P_2 = A(P/A, i, n)$$

$$= \$143.47 \ (P/A, 1.5\%, 116)$$

$$= 143.47 \ (54.8135)$$

$$= \$7864.09$$

The investor therefore is willing to pay only $7864 for the remaining 114 payments in order to earn 18% on the investment.

Example 12.31.   Find *P, A, F,* Given *G, i,* and *n.*

This example illustrates applications of all three arithmetic gradient equations to the same problem situation.

To operate, maintain, and repair a motor grader costs $1000 the first year and increases $500 per year thereafter, so that the second year's costs are $1500, the third year's $2000, and so on, for 5 years. For this example *i* = 8%.

Find:

1.   The present worth of these costs
2.   The equivalent uniform annual cost
3.   The future worth of these costs

Solution:

1. $P = A(P/A, i, n) + G(P/G, i, n)$

$$= 1000(P/A, 8\%, 5) + 500(P/G, 8\%, 5)$$

$$= 1000(3.993) + 500(7.372)$$

$$= \$3993 + \$3686 = \$7679$$

To further illustrate this concept, if $7679 were put into an account drawing 8% interest on the balance left in the account, and $1000 were drawn out at the end of the first year to pay for operation and maintenance, $1500 the second year, and so on, the account would be drawn down to zero at the end of the fifth year.

2. $A = A + G(A/G, i, n)$

$$= \$1000 + 500(1.846)$$

$$= \$1000 + 923$$

$$= \$1923$$

Or, in other words if every year $1923 were paid into an account drawing 8% interest, $1000 could be drawn out the first year, $1500 the second year, and so on, until the fifth year the account balance would be zero.

$$
\begin{aligned}
3. \quad F &= A(F/A, \, i, \, n) \, + \, G(F/G, \, i, \, n) \\
&= \$1000(5.867) \, + \, 500(10.83) \\
&= \$5867 \, + \, \$5415 \\
&= \$11282
\end{aligned}
$$

To look at this solution another way, if at the end of the first year $1000 were borrowed at 8% to pay for operating and maintenance expenses, $1500 borrowed for the second year, and so on, at the end of the fifth year, the amount owed would be $11,282.

13

Applications to Funding
and Feasibility Problems

13.1. Introduction

Often one of the most challenging and rewarding aspects of a proposed construction project is calculating the financial feasibility of the project. Projects that appear impossible to one developer because of lack of ability to calculate the after-tax rate of return may prove quite profitable to another because of an ability to do the necessary calculations. Whether involving conventional financing or some of the more current and innovative methods, at the heart of most financial calculations lies an ability to apply the basic fundamentals of Time Value of Money calculations to the everyday financial problems of development and construction. This chapter deals with typical financing problems and how to solve them through simple step-by-step applications of basic principles. The key is a 4-step process as follows:

1. Gather the data and set up the problem. For the examples in this text the information comes ready made, but in practice the data gathering is often one of the most trying steps, yet one of the most important steps.
2. Analyze the problem. Draw the cash flow diagrams and separate the problem into simple components.
3. Solve each component of the problem and sum the results.
4. Check everything. Check the original data to be sure it is accurate, up-to-date and complete. Check the solution to be sure it includes all costs and incomes and that all costs have negative signs while incomes are

positive. Then eyeball the answer to see if it is consistent with common sense and about where it should be. This procedure will sometimes uncover simple but important errors such as the commonly recurring error of a misplaced decimal point.

13.2   *Choosing between Alternatives*

When comparing the four methods of repaying a loan in Chapter 12, it was seen that they were all equivalent even though the payments and frequency of payments were different. From the viewpoint of the borrower, then, the true cost of a loan lies not in the number of dollars paid back, but rather in the true interest rate and the timing of repayments. The same concept holds true for a lender or an investor. The lender must get interest for his money, and the value of the loan is determined by the interest rate and the timing of the repayments on the loan. An investor usually must choose between several alternatives: "Shall I put my money into one investment with higher risk and higher interest, or shall I put it in another safer investment like an insured bank account with low risk but correspondingly low interest?"

13.3.   *Minimum Acceptable Rate of Return (MARR)*

Many investors decide whether to accept or reject an investment by evaluating the rate of return (i) compared to the risk. Proposals that yield less than an acceptable rate of return are rejected. The lowest rate of return that the investor would consider acceptable is called, plainly enough, the "Minimum Acceptable Rate of Return, or MARR for short. Most people have some reserve funds, where money is "parked" waiting for a better use. These reserve fund "parking places" include savings accounts, money market funds, bonds, and so forth. The return on these funds provides one measure of the investor's MARR. If the insured savings account is yielding 6% interest then the 6% is a basic MARR and a proposed investment must yield at least 6% plus an additional increment for whatever risk is involved plus another increment to compensate for loss of ready availability (liquidity) of the funds.

Other investors don't actually *own* their reserve funds, but consider access and availability of credit as a sufficient backup for rainy days or other special purpose needs. These people use credit cards, signature loans, or other short-term, high-interest money resources for their source of reserve funds. The investor's basic MARR in this case is higher since the cost of taking money out of this reserve pool of funds is much higher, often in the 15 to 20% range.

In any case, the estimated yield on the proposed investment is compared to the MARR and, if the risks are reasonable, those investments with yields greater than the MARR are eligible for acceptance by the investor, to the extent that funds are available.

The following are some typical problems encountered in the construction industry.

13.4   Find the True Interest

Many financing plans include "add-on" interest at an attractive but fictitious rate.

Example 13.1.   Find the True Interest Rate, Given the Add-on Rate and Terms, and Compare Them with the MARR.

Assume that a $100,000 truck–crane is purchased with a down payment of $20,000, and the balance of $80,000 is financed for 36 months at 10% (nominal annual) add-on interest. (a) What is the true rate of interest? (b) Assume the owner's MARR = 15% and compare.

Solution: The cash flow diagram for the financing only is drawn as shown in Figure 13.1.

The lender computes add-on interest as:

$$(\text{Years of time}) \times (\text{Rate}) \times (\text{Principal})$$

a. In this example,

$$\text{Add-on interest} = 3 \times 0.10 \times \$80,000 = \$24,000$$

The lender then divides the total principal plus "interest" by the number of payments to find the monthly payment:

$$\text{Monthly payment} = \frac{\$80,000 + 24,000}{36}$$

$$= \frac{\$104,000}{36}$$

$$= \$2888.89$$

$$P = \$80,000$$

$$A = \$2888.89$$

$$n = 36$$

$$i = ?$$

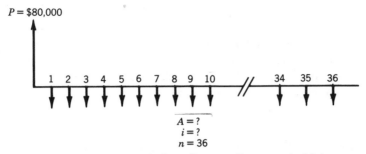

FIGURE 13.1.   Cash flow diagram for example 13.1.

The true rate of interest can be found by solving for i by trial and error.

$$P/A = \$80,000/2888.89 = 27.6923$$

Looking through the tables in the P/A column for $n = 36$ for P/A values on either side of 27.692 yields:

i	P/A
1.5%	27.660
i	27.692
2.0%	25.488

Interpolating gives:

$$i = 1.5\% + \frac{27.692 - 27.660}{27.692 - 25.488} \times (2.0\%) - 1.5\%) = 1.51\%$$

for an annual rate of $(1.0151)^{12} - 1 = 0.197$ or 19.7%/per year

b. Comparison with the owner's MARR of 15%.

An owner should finance the equipment only at interest rates less than the MARR. In this case some inexperienced owners might mistakenly assume that the 10% nominal rate used by the dealer was well below the MARR of 15%. However, the few simple calculations shown above would reveal that the actual interest rate was 19.7%, well above the 15% MARR. The owner should reject this dealer's deal and look elsewhere for better financing.

Example 13.2.   Comparison of the Before-Tax Rates of Return of Different Types of Investments.

Assume that a tax exempt university has funds on hand that either could be placed into a series of short-term bank certificates of deposit yielding 10% interest, compounded annually (representing their MARR), or invested in the purchase of land. The land could be bought for $50,000 now, and in 10 years could probably be sold for $160,000. Property taxes and other expenses are expected to be about $1000 per year. Which is the better investment?

Solution: The cash flow diagrams are drawn as shown in Figure 13.2.

The rate of return, i, for the bank CD is 10%, so the rate of return, i, for the land investment must be found and compared. By trial and error we find the rate of return on the land investment.

				12%	*11%*
Purchase price, $P_1 = $	$-$ $50,000		$= -$$50,000	$-$$50,000
Property tax, etc., $P_2 = $	$-1,000 \times$	$(P/A, i, 10) =$	$-$	5,650 $-$	5,889
		12% 5.6502			
		11% 5.8892			
Resale at EOY 10 $P_3 = $	$+160,000 \times$	$(P/F, i, 10) =$	$+$	51,515 $+$	56,349
		12% 0.32197			
		11% 0.35218			
				$-$$ 4,135 $+$$	460

$$i = 11\% + \frac{460}{460 + 4,135} \times 1\% = 11.1\%$$

In this case the land investment yields 11.1%, just 1.1% above the safer bank certificates. The extra 1.1% yield is probably not sufficient to warrant the extra risk involved in most land investments plus the customary difficulties of trying to sell land on short notice if cash were needed (poor liquidity). A safe, liquid investment in the bank certificates at 10% is probably preferable to a speculative, nonliquid investment in land at 11.1%. Conclusion: Not only should a prospective investment yield more than the MARR, it should also be of comparable risk and liquidity, or else yield a proportionately higher rate of return.

13.5   Equipment Costs before Taxes

Cash in the bank produces not only interest, but also the liquidity necessary to do business. It is a general rule of thumb that liquid assets should not fall below about 10% of the dollar volume of work of a construction com-

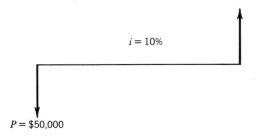

$i = 10\%$

$P = \$50,000$

Investment of $50,000 in bank CDs

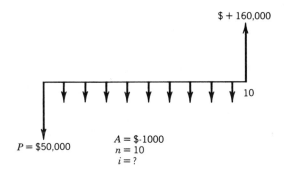

$\$ + 160,000$

10

$P = \$50,000$

$A = \$-1000$
$n = 10$
$i = ?$

Investment of $50,000 in land

FIGURE 13.2. Cash flow diagram for example 13.2.

pany. Some companies have been successful with smaller ratios of cash to dollar volume, but 10% is a generally acceptable figure for a well-managed company.

Example 13.3. Which is Preferable: Cash or Terms?

Assume that the volume of business for your company this year will be ten times the number of dollars of liquid assets owned by your firm and on deposit at the bank. Furthermore, assume that the net profit before taxes will be 5% of the dollar volume of business, and that you earn 7% interest on all liquid asset funds on deposit at the bank.

Suppose that your company wants to buy a piece of equipment that can be purchased for either $100,000 cash, or $20,000 down and the balance at 10% add-on for 3 years. Should you pay cash or buy on terms?

Solution: Before solving the problem in detail, consider what can happen to each $100 that is deposited in the bank.

Year	Money in Bank Beginning of Year	Money in Bank × 10 = Dollars Business	Business × 0.05 = Dollars Profit	Plus 7% Interest on Money in Bank	Accumulated 5% Profit + 7% Interest Money in Bank at End of Year
1	100.00	1000.00	50.00	7.00	157.00
2	157.00	1570.00	78.50	10.99	246.49
3	246.49	2464.90	123.25	17.25	386.99

Notice that, during the first year, the total additional cash that the company has is $50 from profit, and $7 from interest for a total of $57 additional. The profit amounts to $100 × 0.05 = $50, which adds 50% to the initial capital that was in the bank at the beginning of the year. During the second year, the company will make a profit on the original $100, which has not been diminished, and will also make a profit on the interest plus the profit of the first year. It will also receive interest on the first year's profit and the first year's interest. In other words, the profit margin acts like a compound interest rate, provided that it is left in the bank. Therefore, the total "interest" rate becomes $0.07 + (10 \times 0.05) = 0.07 + 0.50 = 0.57$ or 57%.

Now the alternatives become clear. Over a 3-year period, if the cash price of $100,000 for the equipment had been deposited in the bank, the resulting 57% interest compounded would grow to a balance at the end of 3 years of:

$$F = P(F/P, 57\%, 3)$$
$$= \$100,000 (1 + 0.57)^3$$
$$= \$100,000 (3.86989)$$
$$= \$386,989$$

If bought on terms of $20,000 down and 10% add-on interest for 3 years then the monthly payments are calculated as:

Purchase price	$100,000
Down payment	− 20,000
Remaining balance	$ 80,000
Cost of financing 3 years × 0.10/year	× 0.30
Add-on interest charge	$ 24,000

Total to be repaid in monthly installments $80,000 + 24,000 = $104,000.

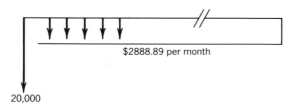

20,000

FIGURE 13.3.   Cash flow diagram for example 13.3.

Monthly Payments $104,000/36 months = $2,888.89 per month

The cash flow diagram for this $20,000 down payment plus $2,888.89 per month plan is drawn in Figure 13.3.

If these funds were deposited in an investment earning 57% interest, the accumulated balance at EOY 3 is calculated below.

The interest on the monthly payments is assumed to be compounded monthly at an interest rate equivalent to a nominal annual rate of 57%. This is calculated as

$$i_e = (1 + i)^{1/n} - 1 = 1.57^{1/12} - 1 = 0.03831$$

$$\text{Down payment } F_1 = P(F/P, 57\%, 3)$$

$$= \$20,000 \times (1.57)^3$$

$$= \$20,000 \, (3.8699)$$

$$= \$77,398$$

$$\text{Monthly payment } F_2 = A(F/A, 3.831\%, 36)$$

$$= A\left[\frac{(1 + i)^n - 1}{i}\right]$$

$$= \$2,888.89 \, \frac{1.03831^{36} - 1}{0.03831}$$

$$= \$2,888.89 \, (74.922)$$

$$= \$216,442$$

If the funds required for buying on terms were invested at 57% per year (or the equivalent 3.831% per month), then the future sum would be

$$F_{\text{total}} = F_1 + F_2$$

$$= \$77,398 + 216,442$$

$$= \$293,840$$

Comparing the future sums of each alternative yields the following tabulation:

Pay $100,000 cash, $F = \$386,989$

Pay $20,000 down and $2,888.89 per month for 36 months

$$F = \$293,840$$

The comparison shows that the equipment should be purchased for $20,000 down and 10% add-on interest, even though the true interest rate (as calculated in Example 13.1) is about double the indicated 10% rate or close to 20%. In effect the company is borrowing $80,000 at about 20% and investing it at 57%, so the numerical answers found above are reasonable.

Many factors were omitted in this problem in order to simplify it—for example, insurance, taxes, maintenance, obsolescence, and depreciation—but in general, the solution indicates that it is best to preserve your capital and use borrowed money to make money, providing the rate of return (i) from the investment is greater than the interest (i) on the borrowed capital. In this case the MARR moves up to 57%, so the borrowing rate is still considerably less than the investment (lending) rate.

13.6. Equipment Costs after Taxes

The application of tax regulations presented herein is intended for illustrative purposes only. Tax laws are subject to change from time to time, and consultation with a qualified tax accountant or lawyer is recommended when doing studies involving taxes.

Income taxes are the major source of revenue for the federal government. Basically income taxes are assessed against each dollar of taxable income as a percentage depending on the tax bracket of the firm or individual earning the income. For instance, a company in the 40% tax bracket would expect to pay 40 cents out of each dollar of taxable income to the U.S. Government for income taxes. The only money the firm gets to keep is the remaining 60 cents out of the taxable $1 earned, and that 60 cents is called "after-tax income." It has been said that the pretax dollar is merely an illusion. What's left after taxes is what counts. However, costs and other expenses involved in earning that $1 of taxable income are deductible from taxable income. Therefore if the firm sold a product for $10 per unit but each unit cost $9 worth of materials, labor, transportation, advertising, and marketing to sell, then the calculations for tax purposes would appear as follows:

Gross income from sales	+$10.00
Costs of production and selling	− 9.00
Taxable income	+$ 1.00
Income tax rate (0.4)	− 0.40
After-tax net income	+$ 0.60

A simplified method of finding after-tax income involves multiplying all costs and incomes by $(1 - t)$ where t represents the tax rate for the company's tax bracket.

For instance, in the example at hand the after-tax income can be found as:

Gross income from sales × $(1 - t)$ = $10.00 × $(1 - 0.4)$ =	$6.00
Cost of production and selling × $(1 - t)$ = $9.00 × $(1 - 0.4)$ =	−$5.40
After-tax income	+$0.60

13.7 Depreciation

When a contractor purchases a new truck for the business for $50,000, is that $50,000 deductible from taxable income? The contractor might say, "Yes, because I spent the $50,000 on equipment used to earn income." The Internal Revenue Service (IRS) would say in effect, "No deduction, because you didn't really spend $50,000, you traded $50,000 worth of cash (or credit) for $50,000 worth of steel, but you still have the $50,000 worth of assets in one form or another, therefore no money has been spent. Now, at the end of a year, when the truck has lost some value due to age and wear, then you can deduct a portion of its value due to depreciation." Capital investments such as the truck, are investments with useful lives greater than 1 year. Depreciable capital investments are depreciated over a number of years called the capital recovery period. They can be depreciated down to zero salvage value either by the Straight Line (SL) method or by an Accelerated Depreciation schedule. The SL method involves simply dividing the purchase price by the years of the capital recovery period and finding the resulting depreciation allowance per year. For instance a truck costing $50,000 new generally can be depreciated to zero salvage value over a 5-year period, thus yielding $50,000 over 5 years, that is, a $10,000 per year depreciation allowance each year for 5 years. If the contractor had $30,000 worth of taxable income before considering depreciation and is in the 40% tax bracket, then the savings due to depreciation are calculated as follows:

		Without Depreciation	With Depreciation
1.	Taxable income before depreciation	$30,000	$30,000
2.	Annual depreciation allowance	none	10,000
3.	Taxable income after depreciation	$30,000	$20,000
4.	Tax-rate bracket	× 0.4	× 0.4
5.	Income tax owed	12,000	8,000
	After-tax income (line 1 minus line 5)	$18,000	$22,000

Notice that $10,000 worth of depreciation raised the after-tax income by $4000 (from $18,000 *without* depreciation to $22,000 *with* depreciation). A quicker and simpler method of calculating the after-tax effects of depreciation is to multiply the depreciation allowance by the tax-bracket rate and count the product as an income resulting from depreciation allowance. In the example problem above, the calculation would be:

$$\text{Depreciation} = \$10,000 \times t$$
$$= \$10,000 \times 0.4$$
$$= \$4000$$

13.8. Investment Tax Credit (ITC)

Around the world the nations that enjoy the highest standards of living are those with the greatest productivity. Productivity is increased by more capital investment in labor-saving equipment and machinery. More productivity results in a more plentiful supply of food and other essential goods at lower prices. Therefore it is in the best interests of most national governments to encourage capital investment in machinery and equipment that produce greater productivity and the resulting prosperity. Governments usually encourage these investments by means of an Investment Tax Credit, or ITC for short. In the United States, the ITC is usually at 10%, but will vary from time to time as Congress tries to fine tune the economy. The ITC consists of a direct deduction from income taxes. For example, if a contractor purchases a $50,000 truck that qualifies for a 10% ITC, a deduction of 10% × $50,000 = $5000 from the income tax due is allowed. The simplest method of dealing with the ITC is to deduct the ITC directly from the purchase price. In the case of the truck, the purchase price of the truck is shown as:

$$\text{Purchase price} \times (1 - \text{ITC}) = \$50,000 \times (1 - 0.1) = \$45,000$$

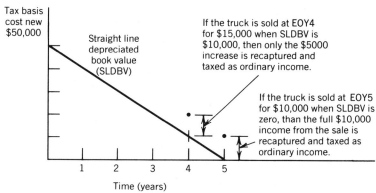

FIGURE 13.4.   Tangible personal property treatment for recapture tax purposes when straight line depreciation is used.

13.9.   Recapture

When personal property (almost all tangible non-real estate property qualifies as personal property) is sold, any gain in excess of the depreciated book value is recaptured by taxing the gain as ordinary income (rather than capital gains), but only to the extent of any prior depreciation, as shown in Figure 13.4. For example, if the $50,000 truck is depreciated to zero book value at the end of 5 years, and then is sold for $10,000, that $10,000 is taxed as ordinary income in the year in which the resale occurred (EOY 5). If the firm is in the 40% tax bracket, it pays $0.40 \times \$10,000 = \$4000$ tax on the income received from the sale of the truck. The amount of after-tax income to the firm is $6000 calculated as ($10,000 \times (1 - t) = \$10,000 \times 0.06 = \$6000$). For two additional examples of recapture, assume (a) the truck is being depreciated at the rate of $10,000 per year (cost new/capital recovery period = $50,000/5 years = $10,000 per year) and is sold at the end of 4 years for $15,000. The depreciated book value at EOY 4 is $10,000, found as:

$$\text{Depreciated book value} = \text{Cost new} - 4 \text{ years} \times \$10,000 \text{ year}$$
$$\text{per year depreciation}$$
$$= \$50,000 - \$40,000$$
$$= \$10,000$$

Then the amount recaptured is $5,000 from:

$$\text{Amount recaptured} = \text{Sales price} - \text{Depreciated book value}$$
$$= \$15,000 - \$10,000$$
$$= \$5000$$

The amount of tax due is $2000 since

$$\text{Recapture tax} = \$5000 \times t$$
$$= \$5000 \times 0.4$$
$$= \$2000$$

For the second example, (b), assume that at the end of the fourth year the truck is sold for $10,000, which is the same value as the depreciated book value. Now there is no recapture since the sale nets no gain in excess of depreciated book value. The sale at this point is tax-free since it constitutes a trade back from steel to cash, for the same reasons the original trade from cash to steel was not deductible in the year of the original purchase.

13.10. Find the After-Tax Cost of Equipment

Example 13.4. Find the equivalent annual after-tax cost of owning and operating a truck, given that a contractor purchases the truck with the data listed below.

Cost		$50,000
Eligible for investment tax credit		10%
Capital recovery period (depreciation life)		5 years
Operating and maintenance (O & M) costs	EOY	O & M
	1	$30,000
	2	35,000
	3	40,000
	4	45,000
	5	50,000

FIGURE 13.5. Cash flow diagram for example 13.4.

Resale value at EOY 5	$10,000
Major overhaul at EOY 3	$15,000
After-tax MARR, $i = 15\%$	
Contractor's tax-bracket rate	40%

Solution: The cash flow diagram is drawn in Figure 13.5.
Separating the problem into simple parts gives the following:

A_1 = Cost new × (1 − ITC) × (A/P, 15%, 5)
  = $50,000 (0.9) (0.29832) = — $13,424

A_2 = Annual depreciation × t = $\dfrac{\$50,000}{5 \text{ years}}$ × 0.40 = +4,000

A_3 = Annual O & M × (1 − t) = $30,000 (0.6) = — 18,000

A_4 = Gradient O & M × (1 − t) × (A/G, 15%, 5)
  = $5000 (0.6) (1.7228) = — 5,168

A_5 = Major overhaul × (1 − t) × (P/F, 15%, 3) (A/P, 15%, 5)
  = 15,000 (0.6) (0.65752) (0.29832) = — 1,765

A_6 = Resale × (1 − t) × (A/F, 15%, 5)
  = $10,000 (0.6) (0.14832) = +890

A_{total} = After-tax cost = — $33,467 per year

The after-tax cost must be covered by before-tax income charges of
($33,467/0.6) = $55,778 per year.

If the truck works 2000 hours per year, then the contractor must charge
the customers $55,778/2000 hours = $27.89 per hour to cover all the costs
and incomes shown and return 15% after taxes on every dollar invested.
If the truck works 1000 hours per year, then the hourly cost becomes
$55,778/1000 hours = $55.78 per hour. If the truck hauls 50,000 cubic
yards of earth per year, then a charge of $55,778/50,000 cubic yard =
$1.12 per cubic yard will cover all cash flows and provide 15% after-tax
return. If the contractor desires to charge an escalating amount with each
future year, any combination of base charge plus gradient can be calculated
as follows.

Example 13.5.   Find G, Given A, i, n.

A contractor needs to receive an equivalent of $27.89 per hour for 2000
hours per year over the 5-year life of the truck used in the previous problem.
However, the contractor wants to charge less now and more in future years.
If the charge starts at $20.00 per hour now, what arithmetic gradient is
needed and what charge rate will be needed at EOY 5?

Solution:

$$A_{total} = \$27.89 \text{ per hour}$$

$$A_1 = 20.00 \text{ per hour}$$

$$G_1 = ?$$

$$i = 15\%$$

$$n = 5 \text{ years}$$

$$A_{total} = A_1 + G_1 \, (A/G, \, 15\%, \, 5)$$

$$27.89 = 20.00 + G_1 \, (1.7228)$$

$$G_1 = \frac{27.89 - 20.00}{1.7228} = \$4.58 \text{ per hour per year}$$

If the rate begins with \$20.00 per hour the first year, the rates will have to increase by \$4.58 per hour each year, so that for year 5 the rates will be \$20.00 per hour + 4 × \$4.58 = \$38.32 per hour

13.11   Application to Real Estate Problems

Problems of real estate financing can be solved with an approach similar to that used for equipment problems. The tax rules are similar but vary in some important details. There is generally no ITC in real estate, and the capital recovery period for depreciation is usually 15 years. Land cannot be depreciated at all, but improvements can be depreciated to zero salvage value. An Accelerated Capital Recovery Schedule (ACRS) is available but a Straight Line (SL) recovery is permissible and will be used here for simplicity. Figures 13.6, 13.7 and 13.8 illustrate some of the IRS rules for treatment of capital gains and recapture of real estate values. Since real estate usually appreciates rather than depreciates, the capital gain rules assume greater importance. Under the 1981 major revision of the tax act, if depreciation of real estate is taken by the SL method, upon resale all gain over the SL-depreciated book value is treated as capital gain. For capital gains, only 40% of the gain is taxed at ordinary income tax rates. If a taxpayer who is in the 30% tax bracket sells a building for \$50,000, and that building has an SL-depreciated book value of \$20,000, then the taxpayer pays a capital gain tax of (sales price − SL-depreciated book value) × 0.40 × tax bracket = (50,000 − 20,000) × 0.4 × 0.3 = \$3,600. An example follows of how to find the rate of return on a proposed investment in a real estate project.

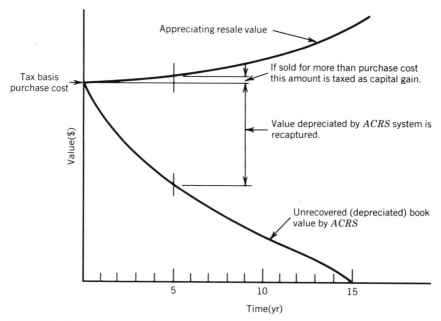

FIGURE 13.6. Capital gains and recapture treatment for *nonresidential* real estate acquired after 1980 and recovered by <u>ACRS.</u>

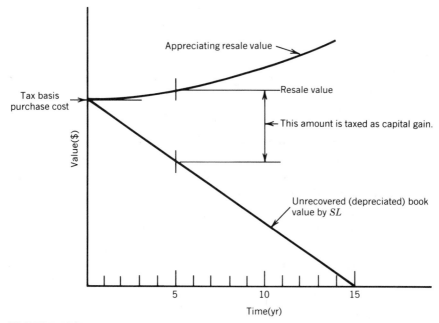

FIGURE 13.7. Capital gains and recapture treatment for *nonresidential* real estate acquired after 1980 and recovered by <u>SL.</u>

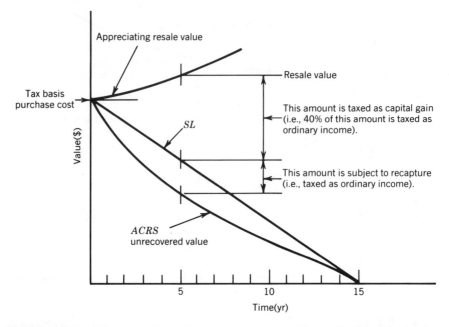

FIGURE 13.8.   Capital gains and recapture treatment for *residential* real estate acquired after 1980.

Example 13.6.   Find the after-tax rate of return (ATROR) on a proposed project outlined as follows:

Land cost: $100,000

Buildings and improvements: $600,000

Capital recovery period: 15 years

Depreciation method: SL

Mortgage: $560,000, 12% fixed rate for 25 years

Interest payments on the mortgage are tax deductible while payments on the principal are not tax deductible (for convenience, assume annual EOY mortgage payments).

O & M costs: $50,000 per year increasing by $4000 per year each year

Rental income: $115,000 per year increasing by $5000 per year each year

The property is expected to appreciate in value by 5% per year and will be sold at EOY 4.

Owner's tax-bracket rate: $t = 30\%$

For simplicity and to avoid repetitious calculations, all periodic costs and rents that are actually incurred monthly are assumed paid at the end of

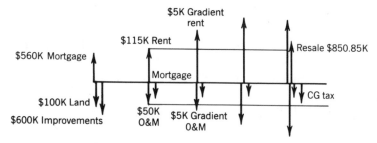

FIGURE 13.9. Cash flow diagram for example 13.6.

each year. As long as monthly rents are higher than monthly costs, the error is conservative and yields a lower than actual ATROR. (For a more detailed analysis of an actual project, the solution can be computerized and costs and incomes calculated on a monthly basis.)

Solution: Following the customary approach, the problem is broken up into small easily digestible bites, solved, and summed. The cash flow diagram is drawn as shown in Figure 13.9.

Then some necessary preliminary calculations are made, as detailed in the following list:

1. How much will the property sell for at EOY 4?

Solution: If it appreciates at 5% per year compounded it will sell for

$$F = P(F/P, 5\%, 4)$$
$$= \$700,00 \ (1.2155)$$
$$= \$850,850$$

2. What is the capital gain tax due on resale at EOY 4?
   a. the annual depreciation allowance =

$$\frac{\text{buildings and improvements}}{\text{capital recovery period}} = \frac{\$600,000}{15 \text{ years}} = \$40,000/\text{year}$$

The SL-depreciated book value (SLBV) =

cost new of land + buildings and improvements −

$$\frac{\text{buildings and improvements}}{\text{capital recovery period}} \times \text{age at sale} = \text{SLBV}$$

$$\$700,000 - \$40,000/\text{year} \times 4 \text{ years} = \$700,000 - \$160,000$$
$$= \$540,000 \text{ SLBV at EOY 4}$$

b) the capital gain at EOY 4 = resale price − SLBV

= $850,850 − 540,000 = $310,850 capital gain at EOY 4

c) the tax on capital gain at EOY 4 = capital gain × 0.40 × tax bracket

= $310,850 × 0.40 × 0.30 = $37,302 capital gain tax at EOY 4

3. A mortgage amortization schedule is needed for the first 4 years, since
    a) interest payments on the mortgage are tax deductible and change
        from year to year, and
    b) the mortgage balance at EOY 4 is needed

$$P = \$560,000$$

$$i = 12\%$$

$$n = 25 \text{ years}$$

The annual payment is calculated as:

$$A = P(A/P, 12\%, 25)$$

$$= \$560,000 \ (0.12750)$$

$$= \$71,400 \text{ per year}$$

Payment for EOY	Payment Amount	Interest	Principal	Mortgage Balance
				$560,000
1	$71,400	$67,200	$4,200	555,800
2	71,400	66,696	4,704	551,096
3	71,400	66,132	5,268	545,828
4	71,400	65,499	5,901	539,927

4. The after-tax rate of return (ATROR) is found by following the ROR
    process developed in the previous chapter. This consists of finding the
    present value (annual or future values could be used just as easily) of
    each component in terms of a trial i. Find one trial i that yields a positive
    sum, and another that yields a negative. Then interpolate to find the
    value of i that yields a zero sum of present worths, and that i is the

rate of return of the project described. For this problem, try a trial i = 20%.

PW using a trial
i = 20%

Cost new at EOY 0

$$P_1 = - \text{ Project cost} + \text{mortgage amount}$$
$$= -100,000 - 600,000 + 560,000 = \qquad -140,000$$

Depreciation credit

$$P_2 = \text{Annual depreciation} \times t \times (P/A, 20\%, 4)$$
$$\qquad\qquad\qquad\qquad\qquad\qquad 2.5887$$
$$= \$40,000/\text{year} \times 0.3 \,(2.5887) = \qquad + \; 31,064$$

Mortgage payments (the interest portion is tax deductible but the principal is not)

$$P_{3-6} = -(\text{interest} \times (1 - t) + \text{principal}) \times$$
$$\qquad\qquad (P/F, 20\ \%, 1 \text{ through } 4)$$
$$P_3 = -(67,200 \times 0.7 + 4200)\,(0.83333) = \qquad - \; 42,700$$
$$P_4 = -(66,696 \times 0.7 + 4704)\,(0.69444) = \qquad - \; 35,688$$
$$P_5 = -(66,132 \times 0.7 + 5268)\,(0.57870) = \qquad - \; 29,838$$
$$P_6 = -(65,499 \times 0.7 + 5901)\,(0.48225) = \qquad - \; 24,957$$

O & M costs

$$P_7 = \text{O \& M} \times (1 - t)(P/A, 20\%, 4)$$
$$= -50,000 \,(0.7)\,(2.5887) = \qquad - \; 90,605$$

Gradient O & M costs

$$P_8 = G \times (1 - t)(P/G, 20\%, 4)$$
$$= - \, 4000 \,(0.7)\,(3.2986) = \qquad - \; 9,236$$

Rental income

$$P_9 = \text{Income} \times (1 - t)(P/A, 20\%, 4)$$
$$= 115,000 \,(0.7)\,(2.5887) = \qquad +208,390$$

Gradient income

$$
\begin{aligned}
P_{10} &= G \times (1 - t)(P/G, 20\%, 4) \\
&= 5000 \, (0.7) \, (3.2986) = \qquad\qquad\qquad + \ 11{,}545
\end{aligned}
$$

Resale net

$$
\begin{aligned}
P_{11} &= (\text{Resale} - \text{Mortgage balance} - \text{Capital} \\
&\quad \text{gains tax}) \, (P/F, 20\%, 4) \\
&= (\$850{,}850 - 539{,}927 \\
&\quad - 37{,}302) \, (0.48225) = \qquad\qquad\qquad + 131{,}954
\end{aligned}
$$

Total PW of the 4-year investment ($i = 20\%$) $ 9,929

The first trial i of 20% yields an after-tax present worth of $+\$9{,}929$, indicating an after-tax rate of return (ATROR) of 20% plus something more. So the PW is recalculated using a second trial i of 25% and a total after-tax present worth of $-\$11{,}709$ is obtained. Interpolating between the two yields an ATROR of $i = 20\% + \dfrac{9{,}929}{9{,}929 + 11{,}709} \times 5\% = 22.3\%$.

Thus, every dollar of the owner's money remaining in the project earns 22.3% compounded annually after taxes.

13.12. Year-to-Year ATROR and Cash Flow

In addition to calculating the ATROR for the entire 4 years, the developer will find it helpful also to determine the cash flow required each year, as well as the year-to-year ATROR. The year-to-year cash flow on real estate is often marginal and not always positive. Particularly in times of rising construction costs and property values, a profitable project may have a negative cash flow right up until the time of resale and then return an attractive profit due to the appreciation in resale price. In the example problem, the ATROR was calculated for a 4-year life. The ATROR for each individual year varies somewhat from the 4-year running average. As time passes, the year-to-year ATROR usually drops, bringing the average ATROR down with it. The drop in ATROR is usually caused by a build-up in equity and the drop in the relative value of the depreciation allowance. The time to sell the project is when the *year-to year* ATROR drops below the owner's minimum acceptable ATROR. If the owner waits until the *average* drops below minimum ATROR, then probably several years of *below*-minimum year-to-year returns have been endured before getting rid of the undesirable project. Calculation of the year-to-year ATROR is illustrated by the following example.

Example 13.7.   Find the Year-to-Year ATROR Given the Data of Example 13.6.

Assume that 4 years ago the owner bought the project described in the previous example. Now, at EOY 4, the owner is deciding whether to keep the project one more year until EOY 5. A year-to-year ATROR figure is needed, so that it can be compared to the owner's minimum acceptable ATROR. Assume all the data used previously is still valid. (In a study of this type, new data can easily be substituted at any time in order to update the study.)

Solution: The same preliminary questions need to be answered, as in the previous example:

1.   How much will the property sell for at EOY 5?

Solution: Still appreciating at 5% per year compounded, one more year compounded at 5% yields

$$\text{Resale at EOY 5} = \text{Resale at EOY 4} \times (F/P, 5\%, 1)$$
$$F = \$850,850\ (1.05)$$
$$= \$893,393$$

2.   What is the capital gain tax due on resale at EOY 5?
   a)   the annual depreciation allowance is still

$$\frac{\$600,000}{15\ \text{years}} = \$40,000\ \text{per year, annual depreciation.}$$

So the SLBV @ EOY 5 =

$$\$700,000 - \$40,000\ \text{per year} \times 5\ \text{years} = \$500,000$$

   b)   the capital gain at EOY 5 = resale price − SLBV =

$$\$893,393 - 500,000 = \$393,393\ \text{capital gain @ EOY 5}$$

   c)   the tax on the capital gain at EOY 5 = Capital gain $\times\ 0.4 \times t$

$$= \$393,393 \times 0.4 \times 0.3 = \$47,207$$

3.   The mortgage amortization schedule is simply a 1-year continuation of the amortization schedule begun in part 3 of the previous example.

$$P = \$560,000$$
$$i = 12\%$$
$$n = 25 \text{ years}$$

Payment for EOY	Payment Amount	Interest	Principal	Mortgage Balance
4				$539,927
5	71,400	64,791	6,609	533,318

4. The calculations for finding the ATROR for 1 year differ somewhat from those used for finding the ATROR for a series of years. In finding the owner's ATROR for any one year, the percentage return for that year is simply the amount of cash–after–taxes the owner can take out at the end of the year divided by the amount of after–tax–cash–equity left in at the beginning of the year (BOY). The amount of cash equity left in at the BOY 5 (same as EOY 4) was found in the previous example but was identified as P_{11} at EOY 4, and was calculated as: Resale price − mortgage balance − capital gains tax = +$850,850 − 539,927 − 37,302 = +$273,621 cash out at EOY 4. The amount of cash after tax that the owner could have taken out of the project is the owner's true equity in the project and is the cash investment upon which year-to-year ATROR is calculated.

EOY 5 After-tax Cash Flow Out of the Project

Annual depreciation credit = Annual depreciation $\times t$
$$= 40,000 \times 0.3 = \qquad\qquad +12,000$$
Annual mortgage payment = Interest $\times (1 - t)$ + Principal =
$$= -64,791 \times 0.7 + 6,609 = \qquad -51,963$$
Annual O & M costs = O & M $\times (1 - t)$
$$= -50,000 \times 0.7 = \qquad\qquad -35,000$$
5th year Gradient, O & M costs = $(n - 1) \times G \times 0.7$ =
$$= -(5 - 1) \times \$4000 \times 0.7 = -11,200$$
Annual rental income = Rental $\times (1 - t)$
$$= 115,000 \times 0.7 = \qquad\qquad +80,500$$
5th year Gradient Rental = $(n - 1) \times G \times 0.7$
$$= (5 - 1) \times 5000 \times .7 = \qquad +14,000$$
EOY 5 Resale cash out = (Resale price − Mortgage balance − CGT)
$$= (+893,393 - 533,318 - 47,207) = +312,868$$

Total after-tax cash out at EOY 5 = $\qquad\qquad$ +$321,205

EOY 4 Cash left in the project = $273,621

$$\text{5th Year, Year-to-Year ATROR} = \frac{\text{After tax cash out at EOY 5}}{\text{Cash left in at EOY 4}} - 1$$

$$= \frac{\$321,205}{\$273,621} - 1 = 0.174 = 17.4\%$$

If the owner's minimum acceptable ATROR is above 17.4%, then the project should be sold at EOY 4 and the funds placed in more desirable projects.

Appendix A

Typical Examples of Forms Required for Obtaining a Development Loan

THE LOAN APPLICATION

APPLICATION FOR LOAN ON RESIDENTIAL PROPERTY
(INCLUDING APPRAISAL REPORT)

Branch _____ Loan Officer _____ Date _____

Loan Desired $_____ Term _____ Rate _____%

PROPERTY DESCRIPTION

Address _____ City _____ State _____

Legal Description _____

Owner of record _____

Lot size _____ X _____ Age ____ Date Purchased _____ Purchase Price $ _____

(include additions or improvements) _____

Proposed Construction: Land Cost $ _____ Estimated Construction Cost $_____

Present Loan: Original Amount $ _____ Balance $ _____ Int. paid to _____

From whom _____ Rate _____ % Date Due _____

Occupied by: Owner_____ Tenant _____ Rented for $_____ per month

APPLICANT'S HISTORY

Name _____ Age [   ] Wife's Name _____

Address _____ Phone Number _____

Employed by _____ For _____ years.   Position _____

Business Address _____ Business Phone Number _____

Annual Income: Salary $_____ Other $ _____ Total $ _____

Source of Income Other Than Salary or Business _____

_____ Credit References _____

_____

Loan Proceeds to be used for _____

Amount collected to apply on Appraisal Fee $ _____

_____

Signature of Applicant

Lending Officer's Comment — Recommendations _____

_____

[Reverse Side]

RESIDENTIAL APPRAISAL REPORT

Street Address _____   LEGAL DESCRIPTION: Lot_____

City _____ State_____   Floors: _____

_____ Side of _____ St.   Block _____

_____ Feet _____ of _____ St.   _____

Paved _____Gravel _____   Addition _____

Walks and Curbs _____   Ground Size _____

Sewer _____ Cesspool _____ Septic Tank_____   Plantings _____
Water Supply_____ Electricity ___ Gas _____   Sprinkling System_____
                                                      Above or Below Grade _____

BUILDING IMPROVEMENTS
   (DESCRIBE):   NEIGHBORHOOD:
Year Built_____ Size _____   Age of typical Property_____
Kind_____ _____   Distance to Schools _____
No. of Stories _____   Distance to Shopping _____
Foundation _____   Distance to Car or Bus _____
Basement Size _____ _____   Zone Restrictions_____
Heating Plan _____   Declining _____
Built-in Equipment _____   Static _____
Water Heater _____   Improving_____
Laundry Trays _____   Quality _____
Recreation Room _____   Adverse Influences _____
Rooms: 1st Floor _____   Walls: _____
Details: 2nd Floor Basement _____

Living Room   _____ _____   Fireplace _____
Dining Room and Halls   _____ _____   Linen Closet _____
Bedrooms   _____ _____   Closets _____
Kitchen   _____ _____   Cabinets _____
Bath   _____ _____   Shower _____
2nd Floor Bedrooms   _____ _____   Closets _____

Plumbing and Light Fixtures: Modern _____   Semi-Modern _____ Old _____
Garage Size _____   Drives and Walks _____
General Condition: Exterior _____   Interior_____ Roof_____
Grounds _____   Improvements Required _____
_____
_____ Cost $ _____

USE FOR INCOME DWELLINGS No. of Rental Units_____ Rooms each Unit _____
   Vacancy Allowance _____ % $ ____   Gross Yearly Rental Furnished $ ____
   Operating Expense $ ____   Gross Yearly Rental Unfurnished $ ____
      Total $ ____   Less Operating and Vacancy $ ____
                                        Net Return $ ____

   ASSESSED VALUE: RENTAL VALUE: APPRAISEMENT:
LAND $ _____   LAND $ _____
IMPROVEMENTS $ _____   IMPROVEMENTS $ _____
TOTAL $ _____ $ _____ TOTAL $ _____

Remarks: _____
Date _____ Appraiser _____

APPLICATION FOR LOAN ON COMMERCIAL PROPERTY
(INCLUDING APPRAISAL REPORT)

Branch _____ Loan Officer _____ Date _____

Applicant _____ Address _____

Loan Desired $ _____ Term _____ _____ Rate _____ %

PROPERTY DESCRIPTION

Address _____ City _____ State _____

Legal Description _____

Land Size _____ x _____ Date Purchased_____ Cost $_____

BUILDING IMPROVEMENTS

Existing _____ Size _____ No. Stories _____ Age _____

To be built _____

Construction _____ Basement Size _____

Date Purchased _____ Purchase Price $ _____

No. of Rentals _____ Annual Gross Income $ _____

Annual Rental Value of Owner Occupied Space $ _____

Annual Operating Cost Including Taxes $ _____

Space Under Lease	Term of Lease	Date Expires

Holder of Present Mortgage _____ Address_____

Original Amt. $ _____ Bal. Due $ _____ Rate _____ % Ins. Amt. $ _____

Financial Statement: Attached [ ]   In File [ ]

_____

Signature of Applicant

Loan Officer's Recommendations _____

_____   _____

_____

[Reverse Side]

APPRAISER'S REPORT

Date _____

Address of Property _____

Legal Description _____

_____

236

THE LOAN APPLICATION

Land Size _____ Size of Bldg. _____

No. of Stories _____ Type of Construction _____ Year Built _____

Foundation _____ Basement Size _____

Heating Plant _____

Plumbing _____ Elevators _____

Fire Escapes _____ Type of Roof _____

Use/Type of Building _____

OCCUPANCY	Gross per Month	Termination of Lease
1st Floor _____	$ _____	_____
2nd Floor _____	$ _____	_____
3rd Floor _____	$ _____	_____
4th Floor _____	$ _____	_____
Additional Space _____	$ _____	_____

Total Gross Annual Income 100% Occupancy $ _____

Less _____ % Predicted Vacancy $ _____

Effective Gross Annual Income Total $ _____

ANNUAL OPERATING EXPENSE

Total General Taxes	$ _____	Repairs – Decorating	$ _____
Special Assessments	$ _____	Equipment Replacement	$ _____
Insurance	$ _____	Supplies	$ _____
Pay Roll	$ _____	Management	$ _____
Heat	$ _____	Advertising	$ _____
Water/Electricity/Gas	$ _____	Miscellaneous	$ _____

Total Operating Expense $ _____

Net Annual Income $ _____

Land value $ _____ Capitalization Rate _____ % $ _____

Annual Net Building Return $ _____

LOCATION INFORMATION

Is the area improving ( ) Static ( ) Declining ( )

Predicted economic life _____ years.

Likelihood of competitive construction: Strong ( ) Moderate ( ) Little ( )

Percentage of occupancy in competitive buildings _____ %

Is the site developed to its highest and best use _____

Remarks: _____

ASSESSED VALUE:		APPRAISEMENT:	
LAND	$ _____	LAND	$ _____
BUILDINGS	$ _____	BUILDINGS	$ _____
TOTAL	$ _____	TOTAL	$ _____

Appraiser

237

PROJECT DATA CHECK LIST FOR
OFFICE BUILDING LOAN

I. Brief Statement of Project
- what type building to be built (e.g., 4 story, 44,000 sq. ft.)
- where building is to be located
- in some instances, statement might conclude with amount only of loan requested.

II. Location and Physical Description
    A. General economic factors of metropolitan area
        1. Brief history of city
        2. City's major attractions
        3. Transportation network
            a. Highways
            b. Airlines
            c. Railroads
            d. Trucking
            e. Bus lines
            f. Water transportation (if applicable)
        4. Trade area
            a. Number of square miles
            b. Population
            c. Buying income
        5. Industry
            a. Number of manufacturing establishments
            b. Number of persons employed in these
            c. Annual payroll
            d. Type products manufactured — are they diversified?
            e. Primary economy of town
        6. Natural resources of area
    B. Population
        1. Rate of growth past five years
        2. Anticipated growth or factors or conditions likely to influence future growth
        3. Breakdown of population
- what percent industrialized, etc.
- number of doctors
- number of dentists
- number of lawyers
- number of insurance companies
- average per capita income

    C. Site location
        1. Brief statement of immediate area around building location
        2. Transportation relative to building
            a. City bus lines nearby — percent of total passengers these lines carry

b. Airport
- time and distance from building
- is it convenient to building occupants/customers
c. Train terminals (in cities where passenger train load is sufficient to warrant discussion)
3. Relation to banks, post office, shops, downtown area
4. Relation to restaurants and other eating facilities
5. Relation to hospitals
6. Relation to predominant residential area for managers, executives, professional people who would be potential occupants.
7. Number of nearby small towns or communities, and time and distance factors relative to building.
8. Outlook for the future for location with particular regard to directional trends of city's present and potential expansion.
9. Existing buildings in area — number, age, height, type, condition, rental rate, and parking facilities
D. Legal description of property
E. Land value
1. Recent sales of comparable property in surrounding area
a. Address
b. Location — relative to proposed building location
c. Dimensions of property
d. Sales price
e. Price per square foot
f. Nature of property, i.e., land, land with improvements taken into consideration, land with improvements not considered.
F. Title to property

III. Proposed Building
A. Detailed description
B. Services (from brochure)
C. Management
D. Type of tenants desired/obtained
E. Description of leases being executed
F. Zoning
G. Taxes
1. City real estate tax — assessment ratio
2. State and county real estate tax
3. Other taxes (where applicable)
a. Franchise and excise tax rates
b. Personalty tax rate
c. Income tax rate
4. Special assessments and charges e.g., garbage pick-up, sewage disposal
H. Insurance rates
I. Average utility costs for building

IV. Estimated Income and Basis of Estimates
  A. Proposed rental rate for building
      ● ground floor
      ● upper floors
      ● comparison with rates of existing buildings — favorable or unfavorable and why
  B. Projections (with comparative operating ratios)
      1. Income statement
      2. Cash flow

V. Economic Feasibility of the Project
  A. Relationship of site to area development
  B. Anticipated factors likely to influence future growth — demand
  C. Results of interviews
      ● Standard set of questions employed
  D. Conclusions

VI. Loan Requested
  A. Amount
  B. Term
  C. Interest rate

VII. Information Concerning Applicant
  A. Company background (including profit-sharing plan)
  B. Principals

VIII. Exhibits
  A. Plans and specifications
  B. City map
  C. Office building brochure
  D. Company brochure
  E. Photographs (aerial and ground) of site
  F. Sample lease
  G. Typical office layout
  H. Pictures of surrounding properties

ANALYSIS OF THE TRANSACTION

ANALYSIS OF DRIVE-IN RESTAURANT SITE

DATE _____   REPORTED BY: _____

I. LOCATION: _____
                (Street address, City, County and State)

II. SITE DATA & ACCESSIBILITY: *(ATTACH PLOT PLAN; PHOTOGRAPH-LEFT SIDE, RIGHT SIDE, DIRECTLY IN FRONT OF, ACROSS STREET FROM; LEGAL DESCRIPTION; CITY MAP WITH LOCATION PINPOINTED.)*

	Yes	No	Comment
Lot is level	[ ]	[ ]	_____
Drainage	[ ]	[ ]	_____
Curbs	[ ]	[ ]	_____
Gutters	[ ]	[ ]	_____
Sidewalk	[ ]	[ ]	_____
Sewer	[ ]	[ ]	_____
Water	[ ]	[ ]	_____
Gas	[ ]	[ ]	_____
Electricity	[ ]	[ ]	_____
Utility poles to be relocated	[ ]	[ ]	_____
Good street condition	[ ]	[ ]	_____
Traffic signal (if at intersection)	[ ]	[ ]	_____
Bus stop nearby	[ ]	[ ]	_____
Offsite parking available	[ ]	[ ]	_____
Proposed street changes or repairs	[ ]	[ ]	_____
Congested traffic	[ ]	[ ]	_____

Character of traffic: neighborhood _____ % Business _____ % Tourist _____ %.
Speed zone of traffic in front of site _____ mph, adjacent street _____ mph.
Site is right-hand _____ or left-hand _____ outbound.
How far back are buildings on adjacent property? (in feet) _____
Visibility distance approaching site street:

    From left: Sign _____ yds.   Bldg. _____ yds.
    From right: Sign _____ yds.   Bldg. _____ yds.

III. FAST FOODS COMPETITION WITHIN ONE MILE OF SITE

Name	Type	Distance from our site
_____	_____	_____
_____	_____	_____
_____	_____	_____

(If more room needed, attach separate sheet.)

Supermarkets in Area: _____

Distance from Site: _____ Hours: _____

Churches in Area: _____

Distance from Site: _____

Department Stores in Area: _____

Distance from Site: _____

Schools in Area: _____

Distance from Site: _____

Shopping Centers within one mile: Yes [ ]   No [ ] .   Size: Major [ ]

Neighborhood [ ]   Small [ ].

IV. POPULATION AND INCOME DATA:

1. Show population by quadrant on Diagram 2. Use (m) for thousands.
2. Indicate income level by quadrant on Diag. 2. Source _____
3. Indicate residential value by quadrant on back. Source _____
4. Age of homes range from _____ yrs. to _____ yrs. Source _____
5. No. of homes within 1 mile _____ ; 2 miles _____
6. Major industry firms in this area are _____ .

V. ZONING DATA:

1. Is property zoned for a restaurant? Yes [ ] ; No [ ] . If "yes", what is classification of zoning? _____ . If "no", rezoning feasibility _____ and time required? _____

2. Front set-back required: _____ ft. Side yard inset must be _____ ft. Rear yard inset must be _____ ft.
   Source of zoning data (give official name & phone number): _____
   _____

3. Our sign permitted? Yes [ ]   No [ ] . If yes, zoning class existing _____
   Maximum Sq. ft. sign area _____ Maximum height on sign _____
   Set-back on sign (in feet) _____

4. How many curb cuts permitted? _____ Maximum width of curb cuts ____ Min. _____
   Traffic count, 24 hour period _____ and _____
   Front of Site (street)   Adjacent (street)

5. Owner _____ Address _____
   City _____ State _____ Phone _____

6. Realtor _____ Address _____
   City _____ State _____ Phone _____
   Attorney _____ Address _____
   City _____ State _____ Phone _____

ANALYSIS OF THE TRANSACTION

7.   Improved Lease _____
     Asking Rental with improvements: _____
     What will the taxes be with improvements: _____
     Will landlord give 15 year lease and options _____ . Will landlord give option to purchase _____

8.   Land Lease _____ Asking Rental _____ Term _____ Tax Rate is _____

9.   Purchase _____ Asking Price _____ Cash _____ Contract _____
     Taxes are: _____ Assessed Valuation: _____

VI. LAND INFORMATION:
     1.   Lot size _____ . Is it a corner lot? _____ Is alley next to property? _____
          Size _____ . Attach plot plan or drawing of property.
     2.   Is land at street level _____ . If not, give facts on _____
     3.   Type of fill used _____ . When filled _____
     4.   Improvements on property are: _____
          _____
     5.   Is land clear of trees and all other conditions regarding site work? _____
          Give facts, size and number of trees, type of fill on lot _____
          _____   _____

DIAGRAM NO. 1
(Show North)

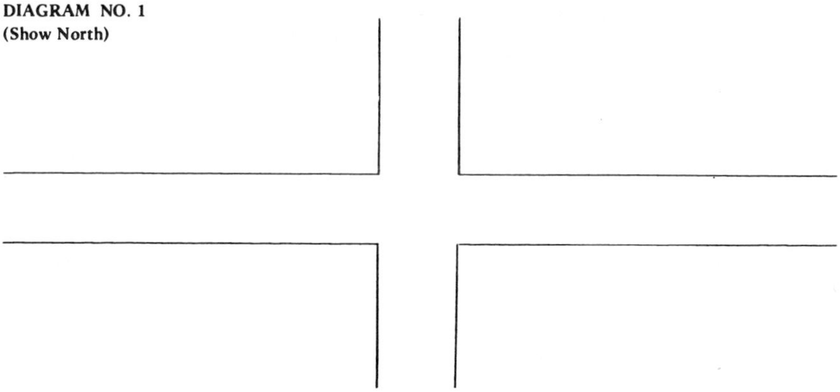

VII. COMMENTS:

_____
_____
_____
_____

243

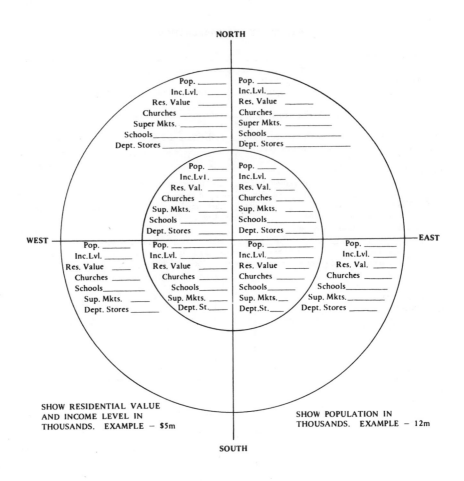

NORTH

Pop. _____	Pop. _____
Inc.Lvl. _____	Inc.Lvl._____
Res. Value _____	Res. Value _____
Churches _____	Churches _____
Super Mkts. _____	Super Mkts. _____
Schools_____	Schools_____
Dept. Stores _____	Dept. Stores _____

Pop. _____	Pop. _____
Inc.Lvl. ___	Inc.Lvl. ___
Res. Val. ____	Res. Val. ____
Churches _____	Churches _____
Sup. Mkts. _____	Sup. Mkts. _____
Schools _____	Schools _____
Dept. Stores _____	Dept. Stores _____

WEST ────

Pop. _____	Pop. _____	Pop. _____	Pop. _____
Inc.Lvl. _____	Inc.Lvl. _____	Inc.Lvl._____	Inc.Lvl. _____
Res. Value ____	Res. Value _____	Res. Value _____	Res. Val. _____
Churches _____	Churches _____	Churches _____	Churches _____
Schools_____	Schools_____	Schools_____	Schools_____
Sup. Mkts. _____	Sup. Mkts. _____	Sup. Mkts. __	Sup. Mkts. _____
Dept. Stores _____	Dept. St._____	Dept.St.____	Dept. Stores _____

──── EAST

SHOW RESIDENTIAL VALUE
AND INCOME LEVEL IN
THOUSANDS. EXAMPLE — $5m

SHOW POPULATION IN
THOUSANDS. EXAMPLE — 12m

SOUTH

────────────────

ANALYSIS OF THE TRANSACTION

DIAGRAM OF PLOT

MUST BE DRAWN IN DETAIL, SHOWING DIRECTIONS AND SIDE OF ROAD: NORTH, SOUTH, EAST OR WEST; PROPOSED LOCATION OF STORE; CURB CUTS OR BREAKS ALREADY ON PROPERTY — GIVE NUMBER OF CURB CUTS PERMITTED: SIDEWALKS AND WHO MAINTAINS. SHOW ANY TREES, BUILDINGS, POLES, FIRE HYDRANTS, TRAFFIC LIGHTS, BUS STOPS. ATTACH TO SURVEY.

THE LOAN COMMITMENT

CHECK LIST OF DOCUMENTS TO BE RECEIVED

BROKER OR SPONSOR _____

[ ] Construction
[ ] Permanent
[ ] Construction & Permanent
[ ] Stand-By

BORROWER _____

LOCATION _____

Date Requested	Documentation	Date Received	Remarks
	Loan Submission:		
	Financial Statement — Borrower		
	Financial Statement — Guarantors		
	Financial Statement — General Contractors		
	D. & B. Report		
	Reference Checks (Personal, Retail, Banks, etc.)		
	Plot Plan		
	Plans and Specifications — Preliminary		
	Cost Breakdown		
	Proposed Annual Income and Operating Statement		
	Appraisal (By: _____)		
	Market Analysis (By: _____)		
	Photographs		
	Take-Out (Who: _____)		
	Lease Agreements		
	Lessee Rated (_____) and Term (_____ yrs.)		
	Committee Brief		
	Approved by Committee		
	Commitment Letter Mailed or Delivered		
	Commitment Letter Accepted		
	Loan Approved:		
	Fee Payable — Cash		
	Fee Payable — Note Due _____ , 19_____		
	Partnership or Trust Agreement		
	Corporate Resolution		
	Corporate Borrowing Authority		
	Corporate Articles		

Date Requested	Documentation	Date Received	Remarks
	Guaranty Agreement		
	Note		
	Mortgage		
	Construction Loan Agreement		
	Hazard Insurance (Builders Risk)		
	Final Policies of Insurance — Premiums Prepaid		
	Liability Insurance		
	Guaranteed Rental Scale, Lease Insurance		
	Title Policy		
	Escrow Letter (Transfer Title Only)		
	Letter of Instructions — Title Company (Disburse)		
	Schedule "B" Title Exceptions Approved by Legal Department		
	Participation Agreement		
	(Who: _____)		
	Buy/Sell Agreement		
	Construction Contract		
	Interest Equalization Tax Agreement		
	Draw Schedule		
	Financing Statements and Security Agreements — Chattels		
	Inventory of Personal Property Identified by Make, Model and Serial Number		
	Financing Statements and Security Agreements — Receivables		
	Evidence That Personal Property, Fixtures and Equipment Are Free and Clear of All Liens and Encumbrances		
	Certificate of Occupancy or Other Government Certificate re: Zoning, Ordinances, etc.		
	Evidence Showing Compliance With Zoning Laws and Ordinances As Of The Completion Date		
	Plans & Specifications — Final		
	Plat of Survey on Construction Completion		
	Architect's Certificates of Completion Per Plans and Specifications		
	Original Building Permit From City, County or Other Governing Authority		

THE LOAN COMMITMENT

Date Requested	Documentation	Date Received	Remarks
	Evidence of Compliance With Government Regulations in Connection With Sale of Electricity, Utilities or Other Unique Functions		
	Verify Accuracy of Interim Interest Charges Through Closing		
	Certificates of Occupancy For All Tenants		
	Tentant Space Acceptances		
	Subordination Agreement		
	Assignment of Mortgage (If Previously Held by Interim Lender)		
	Original Prime Leases as Recorded		
	Lease Assignments (recorded prior to mortgage)		
	Assignment of Assignment of Lessor's In Interest (If previously held by Interim Lender)		
	Compliance With All Lease and Leasing Requirements Prior to Completion Date, Per Borrower's Affidavit		
	Evidence of Compliance with Parking Index of Gross Leaseable Area to be Maintained		
	Establish Tickler Fire Quarterly & Annual Statements re: Kicker		
	Legal Department Memo Conditions Fulfilled		

Financial Statement for Individual Partnership or Corporation

To_____ **Bank**

_____City_____State

Name_____

Address_____   Character of Business_____

For the purpose of obtaining credit from the above Bank, I hereby tender the following statement of my business as of_____

_____ 19_____ and agree to notify you promptly of any change affecting my ability to pay.

Every schedule on this report must be filled out. Where there are no amounts to enter, write the word "None".

ASSETS				LIABILITIES			
Cash in hand and in banks				Notes Payable, due within one year { For merchandise			
Notes receivable, not due, but maturing within six months { For merchandise sold					For real estate		
	For money loaned				To banks		
	For real estate sold				To others...................		
Accounts Receivable, current and considered good { For merchandise sold.. ..				Accounts Payable { For merchandise, not due..			
	Other				For merchandise, past due		
Merchandise (indicate whether cost or replacement value)					Others, not due		
United States Bonds or Notes					Others, past due..........		
Other Listed Securities (Give detailed list on back)				Other current liabilities			
TOTAL QUICK				TOTAL CURRENT			
Notes Receivable, past due or maturing in over six months { For merchandise sold....				Notes Payable, not due within twelve months from date of this statement { Real estate mortgage ..... (List on reverse side)			
	For money loaned........				Other		
	For real estate sold......				Other mortgages, judgments or liens not within twelve months from date of this statement (List in detail)			
Accounts Receivable, past due { For merchandise sold							
	Other							
Real Estate (List on reverse side)				....			
Buildings (List on reverse side)...................				....			
Machinery and tools $....................				TOTAL LIABILITIES			
less depreciation....................net				NET WORTH			
Stocks and bonds other than U. S. (See reverse side)				This section to be used only if a corporation to show distribution of net worth			
Other assets (Itemize)							
....				Capital Stock, common			
....				Capital Stock, preferred			
....				Surplus			
....				Undivided Profits ...:....................			
TOTAL ASSETS.................				TOTAL.................			

Total sales last calendar or fiscal year $_____. Cost of sales $_____. Profit or loss $_____

Have endorsed notes of others amounting to $_____, In present business_____years. Carry $_____

insurance on merchandise $_____on buildings, and $_____on machinery and equipment.

Carry $_____life insurance payable to_____. Have pledged $_____ _____

of above accounts and notes as collateral.

Form No. A-1   OVER

SCHEDULE OF REAL ESTATE OWNED AND MORTGAGES PAYABLE

Location and Description	Improvements	Value		Mortgages	Equity
		Assessed	Cash		

NOTE--If you have ever failed in business give particulars below and how and on what basis you settled with creditors.

_____

_____

_____

_____

_____

Listed Stocks and Bonds other than U. S. Bonds--See opposite side

DESCRIPTION	MARKET VALUE

PLEASE ANSWER FULLY:

1. Are you a partner in any firm?_____Name of firm_____

2. Is real estate as listed recorded in your name?_____

    If not, in whose name?_____

    If joint, state with whom_____

3. If this statement covers the business of a partnership list below the names and addresses of all partners.

_____

_____

I certify that the above schedules and the statements on the opposite side are a true and correct account of the condition of my business and affairs on the day above stated.

Witness my hand and seal, this_____day of_____19_____

_____(Seal)

Appendix B*

A Partial List of Federal Programs Involving Funding for Dwelling Project Construction

*All material in Appendix B is taken from "Programs of HUD," which is updated periodically. Questions concerning program changes should be addressed to Leonard Burchman, Assistant to the Secretary for Public Affairs, Department of Housing and Urban Development, 451 Seventh Street SW, Washington, D.C. 20410.

HUD-FHA PROGRAMS

Title

I   Home Improvement Loans . . . .
             Manufactured Home Loans . . . .

I   Community Development Block Grants . . . .
             (Housing and Community Development
             Act of 1974)

VI   Equal Opportunity in HUD-Assisted . . . .
             Programs
             (Civil Rights Act of 1964)

VIII

             Fair Housing . . . .
             (Civil Rights Act of 1968)

Section

8   Lower-Income Rental Assistance . . . .
             (U.S. Housing Act of 1937)

23   Low-Rent Leased Public Housing . . . .
             (U.S.Housing Act of 1937)

202   Direct Loans for Housing for the . . . .
             Elderly or Handicapped
             (Housing Act of 1959)

203   One-to-Four Family Home Mortgage . . . .
             Insurance
             (National Housing Act)

207   Multifamily Rental Housing . . . .
             (National Housing Act)

213   Cooperative Housing . . . .
             (National Housing Act)

HUD-FHA

253

One- To Four-Family Home Mortgage Insurance (Section 203(b) and (i))

Federal mortgage insurance to facilitate homeownership and the construction and financing of housing.

Nature of Program: By insuring commercial lenders against loss, HUD encourages them to invest capital in the home mortgage market. HUD insures loans made by private financial institutions for up to 97 percent of the property value and for terms of up to 30 years. The loan may finance homes in both urban and rural areas (except farm homes). Less rigid construction standards are permitted in rural areas.

HUD/FHA-insured homeowners threatened with foreclosure due to circumstances beyond their control, such as job loss, death, or illness in the family, may apply for assignment of the mortgage to HUD which, if it accepts assignment, takes over the mortgage and adjusts the mortgage payments for a period of time until the homeowners can resume their financial obligations.

Applicant Eligibility: Any person able to meet the cash investment, the mortgage payments, and credit requirements.

Legal Authority: Section 203, National Housing Act, (P.L. 73-479).

Administering Office: Assistant Secretary for Housing-Federal Housing Commissioner, Department of Housing and Urban Development, Washington, D.C. 20410.

Information Source: HUD Field Offices.

Current Status: Active.

Scope of Program: Cumulative activity through September 1981: 11,212,822 units insured under Section 203(b) for a value of $161,106,408,504; 78,969 units in outlying areas insured under Section 203(i) for a value of $599,508,100.

HUD-FHA

Multifamily Rental Housing (Section 207)

Federal mortgage insurance to facilitate construction or rehabilitation of a broad cross section of rental housing.

Nature of Program: HUD insures mortgages made by private lending institutions to finance the construction or rehabilitation of multifamily rental housing by private or public developers. The project must contain at least five dwelling units. Housing financed under this program, whether in urban or suburban areas, should be able to accommodate families (with or without children) at reasonable rents. The housing project must be located in an area approved by HUD for rental housing and in which market conditions show a need for such housing.

Applicant Eligibility: Investors, builders, developers, and others who meet HUD requirements may apply for funds to an FHA-approved lending institution after conferring with their local HUD office. The housing project must be located in an area approved by HUD for rental housing and in which market conditions show a need for such housing.

Legal Authority: Section 207, National Housing Act, (P.L. 73-479), as amended.

Administering Office: Assistant Secretary for Housing-Federal Housing Commissioner, Department of Housing and Urban Development, Washington, D.C. 20410.

Current Status: Active.

Scope of Program: Cumulative projects insured through September 1981: 2,261 projects with 285,716 units insured with a value of $3,763,141,615.

HUD-FHA

Homeownership Assistance for Low- and Moderate-Income Families (Section 221(d)(2))

Mortgage insurance to increase homeownership opportunities for low- and moderate-income families, especially those displaced by urban renewal.

Nature of Program: HUD insures lenders against loss on mortgage loans to finance the purchase, construction or rehabilitation of low-cost, one- to four-family housing. Maximum insurable loans for an owner-occupant are $31,000 for a single-family home (up to $36,000 in high cost areas). For a larger family (five or more persons), the limits are $36,000 or up to $42,000 in high costs areas. Higher mortgage limits apply to two- to four-family housing.

Applicant Eligibility: Anyone may apply; displaced households qualify for special terms.

Legal Authority: Section 221(d)(2), National Housing Act, (P.L. 73-479).

Administering Office: Assistant Secretary for Housing-Federal Housing Commissioner, Department of Housing and Urban Development, Washington, D.C. 20410.

Information Source: HUD Field Offices.

Current Status: Active.

Scope of Program: Cumulative activity through September 1981: 921,112 units insured for a value of $13,102,587,540.

HUD-FHA

Multifamily Rental Housing for Low- and Moderate-Income Families (Sections 221(d)(3) and (4))

Mortgage insurance to finance rental or cooperative multifamily housing for low- and moderate-income households.

Nature of Program: To help finance construction or substantial rehabilitation of multifamily (5 or more units) rental or cooperative housing for low- and moderate-income or displaced families, HUD conducts two related programs. Both insure project mortgages at the FHA ceiling interest rate. Projects in both cases may consist of detached, semi-detached, row, walk-up, or elevator structures. The insured mortgage amounts are controlled by statutory dollar limits per unit which are intended to assure moderate construction costs. Units financed under both programs may qualify for assistance under Section 8 if occupied by eligible low-income families.

Currently, the principal difference between the programs is that HUD may insure 100 percent of total project cost under Section 221(d)(3) for nonprofit and cooperative mortgagors but only 90 percent under Section 221(d)(4) irrespective of the type of mortgagor.

Formerly, the two programs were distinguished by these additional differences. Projects financed under 221(d)(3) could qualify for a below-market interest rate (as low as 3 percent) and for rent supplements. Consequently, these projects were limited to a lower statutory cost ceiling per unit than was allowed under 221(d)(4) projects which did not benefit from these subsidies. Below-market interest rates and rent supplements are no longer available for new projects for these programs.

Applicant Eligibility: Sections 221(d)(3) and 221(d)(4) mortgages may be obtained by public agencies; nonprofit, limited-dividend or cooperative organizations, private builders or investors who sell completed projects to such organizations. Additionally, Section 221(d)(4) mortgages may be obtained by profit-motivated sponsors. Tenant occupancy is not restricted by income limits, except in the case of tenants receiving subsidies.

Legal Authority: Sections 221(d)(3) and (4), National Housing Act, (P.L. 73-479).

Administering Office: Assistant Secretary for Housing-Federal Housing Commissioner, Department of Housing and Urban Development, Washington, D.C. 20410.

Information Source: HUD Field Offices.

Current Status: Active.

Scope of Program: Cumulative activity through September 1981:
Section 221(d)(3) - 3,465 projects with 349,734 units insured for a value of $5,484,684,629.
Section 221(d)(4) - 5,293 projects with 592,046 units insured for a value of $13,532,790,291.

HUD-FHA

Land Development (Title X)

Federal mortgage insurance to assist development of large subdivisions on a sound economical basis.

Nature of Program: HUD insures mortgages to finance the purchase of land and development of building sites for subdivisions including water and sewer systems, streets and lighting, and other installations needed for residential communities. Community buildings, such as schools, are not included, except for water supply and sewage disposal installations, clubhouses, and parking facilities, owned and maintained jointly by property owners.

Applicant Eligibility: Prospective developers, subject to the approval of HUD, are eligible for mortgage insurance. Public bodies are not eligible.

Legal Authority: Title X of the National Housing Act, amended in 1965 and thereafter; Public Law 89-117.

Administering Office: Assistant Secretary for Housing-Federal Housing Commissioner, Department of Housing and Urban Development, Washington, D.C. 20410.

Information Source: HUD Field Offices.

Current Status: Active.

Scope of Program: Cumulative activity through September 1981: 51 projects with 27,648 lots insured with a value of $187,177,475.

HUD-FHA

Graduated Payment Mortgage (Section 245)

Federal mortgage insurance for Graduated Payment Mortgages.

Nature of Program: HUD insures mortgages to facilitate early homeownership for households that expect their incomes to rise substantially. These "graduated payment" mortgages allow homeowners to make smaller monthly payments initially and to increase their size gradually over time.

Five different payment plans are available, varying in duration and rate of increase. Larger than usual downpayments are required to prevent the total amount of the loan from exceeding the statutory loan to value ratios. In all other respects, the graduated payment mortgage is subject to the rules governing ordinary HUD-insured home loans.

A new Section 245(b) was added by the Housing and Community Development Amendments of 1979, which permits downpayments as low as those allowed under Section 203(b).

Applicant Eligibility: All FHA-approved lenders may make graduated payment mortgages; credit-worthy applicants with reasonable expectations of increasing income may qualify for such loans. The 245(b) program limits eligibility of mortgages to persons who have not owned a home within the past three years.

Legal Authority: Section 245, National Housing Act (1934), (P.L. 73-479).

Administering Office: Assistant Secretary for Housing-Federal Housing Commissioner, Department of Housing and Urban Development, Washington, D.C. 20410.

Information Source: HUD Field Offices.

Current Status: Active.

Scope of Program: Cumulative activity through September 1981: 291,860 units insured with a value of $14,190,454,191.

HUD-FHA

GNMA Mortgage Purchase ("Tandem") Programs

The Government National Mortgage Association (GNMA) creates a secondary mortgage market by purchases of mortgages from private lenders for the purpose of expanding and facilitating investment in housing.

Nature of Program: The Government National Mortgage Association purchases certain types of mortgages to fulfill two statutory objectives: (1) to provide special assistance for the financing of selected types of housing for which financing is not readily available, such as housing for low-income families and (2) to counter declines in mortgage lending and housing construction.
   The multifamily mortgages purchased by GNMA, which generally bear below-market interest rates, are resold at a discount in the market in order to provide acceptable yields to investors with the Government absorbing the loss as a subsidy.

Applicant Eligibility: FHA-approved mortgagees may apply to sell federally underwritten mortgages to GNMA.

Legal Authority: Title III of the National Housing Act, (P.L. 73-479).

Administering Office: Government National Mortgage Association, Department of Housing and Urban Development, Washington, D.C. 20410.

Information Source: Regional officies of the Federal National Mortgage Association in Atlanta, Chicago, Dallas, Los Angeles, and Philadelphia. Also see Administering Office.

Current Status: Full amount of fiscal year 1982 commitment authority has been expended. Mortgages are being purchased pursuant to outstanding commitments issued in fiscal years 1979, 1980 and 1981. Authority to purchase conventional loans lapsed October 1, 1981.

Scope of Program: From fiscal year 1975 through 1982 GNMA issued over $33.6 billion in commitments to purchase federally underwritten and convention: below-market interest rate mortgages under its purchase programs.

HUD-FHA

GNMA Mortgage-Backed Securities Program

Provides a means of channeling funds from the Nation's securities markets into the residential mortgage market.

Nature of Program: GNMA guarantees the timely payment of principal and interest on securities issued by private lenders and backed by pools of Government-underwritten residential mortgages. The program's purpose is to attract non-traditional investors into the residential mortgage market by offering them a high-yield, risk-free, Government-guaranteed security which has none of the servicing obligations associated with a mortgage loan portfolio.

Applicant Eligibility: Applicants must be FHA-approved mortgagees in good standing and have a net worth that meets GNMA's minimum requirements.

Legal Authority: Title III, National Housing Act, (P.L. 73-479), which authorizes GNMA to guarantee the timely payment of principal and interest on the mortgage-backed securities.

Administering Office: Government National Mortgage Association, Department of Housing and Urban Development, Washington, D.C. 20410.

Information Source: Administering office.

Current Status: Active.

Scope of Program: Since its inception, GNMA has guaranteed over $135 billion in mortgage-backed securities, and the program has helped to finance over 3.8 million housing units.

**Major Federal Legislation and
Executive Orders
Authorizing HUD Programs**

(In Chronological Order)

National Housing Act, 1934 (Public Law 73-479)

Title I: Property Improvements
  Section 2: Mobile Homes (Loan Insurance)
    Property Improvement (Loan Insurance)

Title II:
  Section 203: Homes (One-to-Four-Family) (Mortgage Insurance)
  Section 203(h): Disaster Housing
  Section 203(i): Outlying area properties
  Section 203(k): Major Home Improvements (Loan Insurance)
  Section 207: Mobile Home Parks (Mortgage Insurance)
  Section 213: Cooperative Housing (Mortgage Insurance)
  Section 221(h): Major Home Improvements (Loan Insurance)
  Section 221(d)(2): Homes for Low- and Moderate-Income Families
  (Mortgage Insurance)
  Section 221(d)(4): Rental Housing (Market Interest Rate) for
    Low- and Moderate-Income Families (Mortgage Insurance)
  Section 222: Homes for Servicemen (Mortgage Insurance)
  Section 223(e): Housing in Declining Neighborhoods
    (Mortgage Insurance)
  Section 231: Senior Citizen Housing (Mortgage Insurance)
  Section 232: Nursing Homes and Intermediate Care Facilities
    (Mortgage Insurance)
  Section 233: Experimental Housing (Mortgage Insurance)
  Section 234: Condominium Housing (Mortgage Insurance)
  Section 235: Interest Supplements on Home Mortgages
  Section 236: Interest Supplements on Rental and Cooperative
    Housing Mortgages
  Section 237: Mortgage Credit Assistance for Homeownership
  Counseling Assistance for Low- and Moderate-Income
    Families
  Section 240: Purchase of Fee Simple Title from Lessors (Mortgage Insurance)
  Section 241: Insured Supplemental Loans on Multifamily
    Housing Projects
  Section 242: Nonprofit Hospitals (Mortgage Insurance)
  Section 245: Graduated Payment Mortgages

Title III: Government National Mortgage Association

Title VIII:
  Section 809: Armed Services Housing for Civilian Employees
    (Mortgage Insurance)
  Section 810: Armed Services Housing in Impacted Areas
    (Mortgage Insurance)

Title X: Land Development and New Communities (Mortgage Insurance)

Title XI: Group Practices Facilities (Mortgage Insurance)

U.S. Housing Act of 1937 (P.L. 93-383 which replaced P.L. 75-412)

Housing Act of 1949 (P.L. 81-171)

Title I: Urban Renewal Projects

Housing Act of 1950 (P.L. 81-475)

Title IV: College Housing

Housing Act of 1954 (P.L. 83-560)

Title VII:
  Section 701: Comprehensive Planning Assistance

Housing Act of 1959 (P.L. 86-372)

Title II: Section 202: Senior Citizen Housing (Direct Loans)

Housing Act of 1964 (P.L. 88-560)

Title III:
  Section 312: Rehabilitation Loans

Title VIII:
Part 1: Federal-State Training Programs

Housing and Urban Development Act of 1965 (P.L. 89-117)

Department of Housing and Urban Development Act (P.L. 89-174)

Title I: Rent Supplements

Title VII:
  Section 702: Public Water and Sewer Facilities
  Section 703: Neighborhood Facilities

Demonstration Cities and Metropolitan Development Act of 1966 (P.L. 89-754)
Title I: Model Cities
Title X:
  Sections 1010 and 1011: Urban Research and Technology

Civil Rights Act of 1968 (P.L. 90-284)

Title VIII: (Fair Housing)

Housing and Urban Development Act of 1968 (P.L. 90-448)

Title I: Homeownership for Lower-Income Families
Title IV: New Communities
Title VIII: Government National Mortgage Association
Title XI: Urban Property Protection and Reinsurance
Title XIII: Flood Insurance
Title XIV: Interstate Land Sales

Housing and Urban Development Act of 1969 (P.L. 91-152)

Housing and Urban Development Act of 1970 (P.L. 91-609)

Title V: Research and Technology
Title VI: Crime Insurance
Title VII: Urban Growth and New Community Development

Housing and Community Development Act of 1974 (P.L. 93-383)

Title I: Community Development Block Grants
Title II: Assisted Housing
  Section 8: Lower Income Rental Assistance
Title III: Mortgage Credit Assistance
  Section 306: Compensation for Substantial Defects
  Section 307: Coinsurance
  Section 308: Experimental Financing
Title VI: Mobile Home Construction and Safety Standards
Title VIII: Miscellaneous
  Section 802: State Housing Finance Agency Coinsurance
  Section 809: National Institute of Building Sciences (NIBS)
  Section 810: Urban Homesteading
  Section 811: Counseling

Emergency Home Purchase Assistance Act of 1974 (P.L. 93-449)

Emergency Homeowners' Relief Act (P.L. 94-50)
Title I: Standby Authority to Prevent Mortgage Foreclosure

Housing and Community Development Act of 1977 (P.L. 95-128)
  Title I: Community Development
  Title II: Housing Assistance and Related Programs
  Title III: Federal Housing Administration Mortgage Insurance and Related
    Programs
  Title IV: Lending Powers of Federal Savings and
    Loan Associations; Secondary Market Authorities
  Title V: Rural Housing
  Title VI: National Urban Policy
  Title VII: Flood and Riot Insurance
  Title VIII: Community Reinvestment
  Title IX: Miscellaneous Provisions

Housing and Community Development Amendments of 1978 (P.L. 95-557)
Title I: Community and Neighborhood Development and Conservation
Title II: Housing Assistance Programs
Title III: Program Amendments and Extensions
Title IV: Congregate Services
Title V: Rural Housing
Title VI: Neighborhood Reinvestment Corporation
Title VII: Neighborhood Self-Help Development
Title VIII: Livable Cities
Title IX: Miscellaneous

Housing and Community Development Amendments of 1979 (P.L. 96-153)

I. Community and Neighborhood Development and Conservation
II. Housing Assistance Programs
III. Program Amendments and Extensions
IV. Interstate Land Sales
V. Rural Housing
VI. Crime Insurance, Riot Reinsurance, and Flood Insurance Extensions

Housing and Community Development Act of 1980 (P.L. 96-399).

Title I: Community and Neighborhood Development and Conservation
Title II: Housing Assistance Programs
Title III: Program Amendments and Extentions
Title IV: Planning Assistance
Title V: Rural Housing
Title VI: Condominium and Cooperative Conversion Protection and Abuse Relief

Housing and Community Development Amendments of 1981; Title III of the Omnibus Budget Reconciliation Act of 1981 (P.L. 97-35).

Part 1: Community and Economic Development
Part 2: Housing Assistance Programs
Part 3: Program Amendments and Extensions
Part 4: Flood, Crime, and Riot Insurance
Part 5: Rural Housing
Part 6: Multifamily Mortgage Foreclosure
Part 7: Effective Date

265

HUD Regional Offices

Region I
John F. Kennedy Federal Building
Boston, Massachusetts 02203
Area Offices:
Boston, Mass.; Hartford, Conn.
Service Offices:
Manchester, N.H.; Providence, R.I.

Region II
26 Federal Plaza
New York, New York 10278
Area Offices:
New York, N.Y.; Newark, N.J.;
Buffalo, N.Y.; Caribbean
Service Offices:
Albany, N.Y.; Camden, N.J.

Region III
Curtis Building
6th and Walnut Streets
Philadelphia, Pennsylvania 19106
Area Offices:
Pittsburgh, Pa.; Philadelphia, Pa.;
District of Columbia; Baltimore, Md.;
Richmond, Va.
Service Office:
Charleston, W. Va.

Region IV
Richard B. Russell Federal Building
75 Spring Street, S.W.
Atlanta, Georgia 30303
Area Office:
Birmingham, Ala.; Jacksonville, Fla.;
Atlanta, Ga.; Louisville, Ky.;
Jackson, Miss.; Greensboro, N.C.;
Columbia, S.C.; Knoxville, Tenn.
Service Offices:
Coral Gables, Fla.; Tampa, Fla.;
Orlando, Fla.; Memphis, Tenn.;
Nashville, Tenn.

Region V
330 South Wacker Drive
Chicago, Illinois 60606
Area Offices:
Detroit, Mich.; Chicago, Ill.;
Indianapolis, Ind.; Minneapolis, Minn.;
Columbus, Ohio,; Milwaukee, Wisc.
Service Offices:
Cincinnati, Ohio; Cleveland, Ohio;
Grand Rapids, Mich.; Flint, Mich.

Region VI
221 W. Lancaster Ave.,
Fort Worth, Texas 76113
Area Offices:
Dallas, Tex.; Oklahoma City, Okla.;
San Antonio, Tex.; New Orleans, La.;
Little Rock, Ark.
Service Offices: Ft. Worth, Tex.;
Houston, Tex.; Lubbock, Tex.;
Albuquerque, N. Mex.; El Paso, Tex.;
Shreveport, La.; Tulsa, Okla.

Region VII
Professional Bldg.
1103 Grand Ave.
Kansas City, Missouri 64106
Area Offices:
Kansas City, Mo.; St. Louis, Mo.;
Omaha, Nebr.;
Service Office:
Des Moines, Iowa

Region VIII
Executive Tower Bldg.
1405 Curtis Street
Denver, Colorado 80202
Area Office:
Denver, Colo.
Service Offices:
Helena, Mont.; Salt Lake City, Utah

Region IX
450 Golden Gate Avenue
P.O. Box 36003
San Francisco, California 94102
Area Offices:
San Francisco, Calif.;
Los Angeles, Calif.;
Honolulu, Hawaii
Service Offices:
Santa Ana, Calif.; San Diego, Calif.;
Phoenix, Ariz.; Tucson, Ariz.;
Fresno, Calif.; Sacramento, Calif.;
Reno, Nev.; Las Vegas, Nev.

Region X
3003 Arcade Plaza Building
1321 Second Avenue
Seattle, Washington 98101
Area Offices:
Seattle, Wash.; Portland, Oreg.;
Anchorage, Alaska
Service Offices:
Boise, Idaho; Spokane, Wash.

Appendix C*

Interest Tables

| | | $i = 0.5\%$ | | | $i = 0.5\%$ | | | $i = 0.5\%$ | | |
| | PRESENT SUM, P | | | UNIFORM SERIES, A | | | FUTURE SUM, F | | | |
n	P/F	P/A	P/G	A/F	A/P	A/G	F/P	F/A	F/G	n
1	0.99502	0.9950	0.0000	1.00000	1.00500	0.0000	1.0050	1.0000	0.0000	1
2	0.99007	1.9851	0.9900	0.49875	0.50375	0.4987	1.0100	2.0050	1.0000	2
3	0.98515	2.9702	2.9603	0.33167	0.33667	0.9966	1.0150	3.0150	3.0050	3
4	0.98025	3.9505	5.9011	0.24813	0.25313	1.4937	1.0201	4.0301	6.0200	4
5	0.97537	4.9258	9.8026	0.19801	0.20301	1.9900	1.0252	5.0502	10.050	5
6	0.97052	5.8963	14.655	0.16460	0.16960	2.4854	1.0303	6.0755	15.100	6
7	0.96569	6.8620	20.449	0.14073	0.14573	2.9800	1.0355	7.1058	21.175	7
8	0.96089	7.8229	27.175	0.12283	0.12783	3.4738	1.0407	8.1414	28.281	8
9	0.95610	8.7790	34.824	0.10891	0.11391	3.9667	1.0459	9.1821	36.423	9
10	0.95135	9.7304	43.386	0.09777	0.10277	4.4588	1.0511	10.228	45.605	10
11	0.94661	10.677	52.852	0.08866	0.09366	4.9501	1.0564	11.279	55.833	11
12	0.94191	11.618	63.213	0.08107	0.08607	5.4405	1.0616	12.335	67.112	12
13	0.93722	12.556	74.460	0.07464	0.07964	5.9301	1.0669	13.397	79.448	13
14	0.93256	13.488	86.583	0.06914	0.07414	6.4189	1.0723	14.464	92.845	14
15	0.92792	14.416	99.574	0.06436	0.06936	6.9069	1.0776	15.536	107.31	15
16	0.92330	15.339	113.42	0.06019	0.06519	7.3940	1.0830	16.614	122.84	16
17	0.91871	16.258	128.12	0.05651	0.06151	7.8803	1.0884	17.697	139.46	17
18	0.91414	17.172	143.66	0.05323	0.05823	8.3657	1.0939	18.785	157.15	18
19	0.90959	18.082	150.03	0.05030	0.05530	8.8504	1.0994	19.879	175.94	19
20	0.90506	18.987	177.23	0.04767	0.05267	9.3341	1.1049	20.979	195.82	20
21	0.90056	19.888	195.24	0.04528	0.05028	9.8171	1.1104	22.084	216.80	21
22	0.89608	20.784	214.06	0.04311	0.04811	10.299	1.1159	23.194	238.88	22
23	0.89162	21.675	233.67	0.04113	0.04613	10.780	1.1215	24.310	262.08	23
24	0.88719	22.562	254.08	0.03932	0.04432	11.261	1.1271	25.432	286.39	24
25	0.88277	23.445	275.26	0.03765	0.04265	11.740	1.1328	26.559	311.82	25
26	0.87838	24.324	297.22	0.03611	0.04111	12.219	1.1384	27.691	338.38	26
27	0.87401	25.198	319.95	0.03469	0.03969	12.697	1.1441	28.830	366.07	27
28	0.86966	26.067	343.43	0.03336	0.03836	13.174	1.1498	29.974	394.90	28
29	0.86533	26.933	367.66	0.03213	0.03713	13.651	1.1556	31.124	424.87	29
30	0.86103	27.794	392.63	0.03098	0.03598	14.126	1.1614	32.280	456.00	30
32	0.85248	29.503	444.76	0.02889	0.03389	15.075	1.1730	34.608	521.72	32
34	0.84402	31.195	499.75	0.02706	0.03206	16.020	1.1848	36.960	592.11	34
36	0.83564	32.871	557.56	0.02542	0.03042	16.962	1.1966	39.336	667.22	36
48	0.78710	42.580	957.91	0.01849	0.02349	22.543	1.2704	54.097	1219.5	48
60	0.74137	51.725	1448.6	0.01433	0.01933	28.006	1.3488	69.770	1954.0	60
120	0.54963	90.073	4823.5	0.00610	0.01110	53.550	1.8194	163.87	8775.8	120
180	0.40748	118.50	9031.3	0.00344	0.00844	76.211	2.4540	290.81	22163.	180
240	0.30210	139.58	13415.	0.00216	0.00716	96.113	3.3102	462.04	44408.	240
300	0.22397	155.20	17603.	0.00144	0.00644	113.41	4.4649	692.99	78598.	300
360	0.16604	166.79	21403.	0.00100	0.00600	128.32	6.0225	1004.5	128903.	360
INF	0.00000	200.00	43000.	0.0000	0.00500	200.00	INF	INF	INF	INF

	i = 0.75%			i = 0.75%			i = 0.75%			
	PRESENT SUM, P			UNIFORM SERIES, A			FUTURE SUM, F			
n	P/F	P/A	P/G	A/F	A/P	A/G	F/P	F/A	F/G	n
1	0.99256	0.9925	0.0000	1.00000	1.00750	0.0000	1.0075	1.0000	0.0000	1
2	0.98517	1.9777	0.9851	0.49813	0.50563	0.4981	1.0150	2.0075	1.0000	2
3	0.97783	2.9955	2.9408	0.33085	0.33835	0.9950	1.0226	3.0225	3.0075	3
4	0.97055	3.9261	5.8525	0.24721	0.25471	1.4906	1.0303	4.0452	6.0300	4
5	0.96333	4.8894	9.7058	0.19702	0.20452	1.9850	1.0380	5.0755	10.075	5
6	0.95616	5.8456	14.486	0.16357	0.17107	2.4782	1.0458	6.1136	15.150	6
7	0.94904	6.7946	20.180	0.13967	0.14717	2.9701	1.0537	7.1594	21.264	7
8	0.94198	7.7366	26.774	0.12176	0.12926	3.4607	1.0616	8.2131	28.424	8
9	0.93496	8.6715	34.254	0.10782	0.11532	3.9501	1.0695	9.2747	36.637	9
10	0.92800	9.5995	42.606	0.09657	0.10417	4.4383	1.0775	10.344	45.911	10
11	0.92109	10.520	51.817	0.08755	0.09505	4.9252	1.0856	11.421	56.256	11
12	0.91424	11.434	61.874	0.07995	0.08745	5.4109	1.0938	12.507	67.678	12
13	0.90743	12.342	72.763	0.07352	0.08102	5.8954	1.1020	13.601	80.185	13
14	0.90068	13.243	84.472	0.06801	0.07551	6.3786	1.1102	14.703	93.787	14
15	0.89397	14.137	96.987	0.06324	0.07074	6.8605	1.1186	15.813	108.49	15
16	0.88732	15.024	110.29	0.05906	0.06656	7.3412	1.1269	16.932	124.30	16
17	0.88071	15.905	124.38	0.05537	0.06287	7.8207	1.1354	18.059	141.23	17
18	0.87416	16.779	139.24	0.05210	0.05960	8.2989	1.1439	19.194	159.29	18
19	0.86765	17.646	154.86	0.04917	0.05667	8.7759	1.1525	20.338	178.49	19
20	0.86119	18.508	171.23	0.04653	0.05403	9.2516	1.1611	21.491	198.82	20
21	0.85478	19.362	188.32	0.04415	0.05165	9.7261	1.1698	22.652	220.32	21
22	0.84842	20.211	206.14	0.04198	0.04948	10.199	1.1786	23.802	242.97	22
23	0.84210	21.053	224.66	0.04000	0.04750	10.671	1.1875	25.001	266.79	23
24	0.83583	21.889	243.89	0.03818	0.04568	11.142	1.1964	26.188	291.79	24
25	0.82961	22.718	263.80	0.03652	0.04402	11.611	1.2053	27.384	317.98	25
26	0.82343	23.542	284.38	0.03498	0.04248	12.080	1.2144	28.590	345.36	26
27	0.81730	24.359	305.63	0.03355	0.04105	12.547	1.2235	29.804	373.96	27
28	0.81122	25.170	327.54	0.03223	0.03973	13.012	1.2327	31.028	403.76	28
29	0.80518	25.975	350.08	0.03100	0.03850	13.477	1.2419	32.260	434.79	29
30	0.79919	26.775	373.26	0.02985	0.03735	13.940	1.2512	33.502	467.05	30
32	0.78733	28.355	421.46	0.02777	0.03527	14.863	1.2701	36.014	535.31	32
34	0.77565	29.912	472.07	0.02593	0.03343	15.781	1.2892	38.564	608.61	34
36	0.76415	31.446	524.99	0.02430	0.03180	16.694	1.3086	41.152	687.02	36
48	0.69961	40.184	886.84	0.01739	0.02489	22.069	1.4314	57.520	1269.4	48
60	0.63870	48.173	1313.5	0.01326	0.02076	27.266	1.5656	75.424	2056.5	60
120	0.40794	78.941	3998.5	0.00517	0.01267	50.652	2.4513	193.51	9801.9	120
180	0.26055	98.593	6892.6	0.00264	0.01014	69.909	3.8380	378.40	26454.	180
240	0.16641	111.14	9494.1	0.00150	0.00900	85.421	6.0091	667.88	57051.	240
300	0.10629	119.16	11636.	0.00089	0.00839	97.654	9.4091	1121.1	109483.	300
360	0.06789	124.28	13312.	0.00055	0.00805	107.11	14.730	1830.7	196099.	360
INF	0.00000	133.33	17777.	0.0000	0.00750	133.33	INF	INF	INF	INF

	$i=1\%$			$i=1\%$			$i=1\%$			
	PRESENT SUM, P			UNIFORM SERIES, A			FUTURE SUM, F			
n	P/F	P/A	P/G	A/F	A/P	A/G	F/P	F/A	F/G	n
1	0.99010	0.9901	0.0000	1.00000	1.01000	0.0000	1.0100	1.0000	0.0000	1
2	0.98030	1.9704	0.9803	0.49751	0.50751	0.4975	1.0201	2.0100	1.0000	2
3	0.97059	2.9409	2.9214	0.33002	0.34002	0.9933	1.0303	3.0301	3.0100	3
4	0.96098	3.9019	5.8044	0.24628	0.25628	1.4875	1.0406	4.0604	6.0401	4
5	0.95147	4.8534	9.6102	0.19604	0.20604	1.9801	1.0510	5.1010	10.100	5
6	0.94205	5.7954	14.320	0.16255	0.17255	2.4709	1.0615	6.1520	15.201	6
7	0.93272	6.7281	19.916	0.13863	0.14863	2.9602	1.0721	7.2135	21.353	7
8	0.92348	7.6516	25.381	0.12069	0.13069	3.4477	1.0828	8.2856	28.567	8
9	0.91434	8.5660	33.695	0.10674	0.11674	3.9336	1.0936	9.3685	36.852	9
10	0.90529	9.4713	41.843	0.09558	0.10558	4.4179	1.1046	10.462	46.221	10
11	0.89632	10.367	50.806	0.08645	0.09645	4.9005	1.1156	11.566	56.683	11
12	0.88745	11.255	60.568	0.07885	0.08885	5.3814	1.1268	12.682	68.250	12
13	0.87866	12.133	71.112	0.07241	0.08241	5.8607	1.1380	13.809	80.932	13
14	0.86996	13.003	82.422	0.06690	0.07690	6.3383	1.1494	14.947	94.742	14
15	0.86135	13.865	94.481	0.06212	0.07212	6.8143	1.1609	16.096	109.69	15
16	0.85282	14.717	107.27	0.05794	0.06794	7.2886	1.1725	17.257	125.78	16
17	0.84438	15.562	120.78	0.05426	0.06426	7.7613	1.1843	18.430	143.04	17
18	0.83602	16.398	134.99	0.05098	0.06098	8.2323	1.1961	19.614	161.47	18
19	0.82774	17.226	149.89	0.04805	0.05805	8.7016	1.2081	20.810	181.09	19
20	0.81954	18.045	165.46	0.04542	0.05542	9.1693	1.2201	22.019	201.90	20
21	0.81143	18.857	181.69	0.04303	0.05303	9.6354	1.2323	23.239	223.91	21
22	0.80340	19.660	198.56	0.04086	0.05086	10.099	1.2447	24.471	247.15	22
23	0.79544	20.455	216.06	0.03889	0.04889	10.562	1.2571	25.716	271.63	23
24	0.78757	21.243	234.18	0.03707	0.04707	11.023	1.2697	26.973	297.34	24
25	0.77977	22.023	252.89	0.03541	0.04541	11.483	1.2824	28.243	324.32	25
26	0.77205	22.795	272.19	0.03387	0.04387	11.940	1.2952	29.525	352.56	26
27	0.76440	23.559	292.07	0.03245	0.04245	12.397	1.3082	30.820	382.08	27
28	0.75684	24.316	312.50	0.03112	0.04112	12.851	1.3213	32.129	412.91	28
29	0.74934	25.065	333.48	0.02990	0.03990	13.304	1.3345	33.450	445.03	29
30	0.74192	25.807	355.00	0.02875	0.03875	13.755	1.3478	34.784	478.48	30
32	0.72730	27.269	399.58	0.02667	0.03667	14.653	1.3749	37.494	549.40	32
34	0.71297	28.702	446.15	0.02484	0.03484	15.544	1.4025	40.257	625.77	34
36	0.69892	30.107	494.62	0.02321	0.03321	16.428	1.4307	43.076	707.68	36
48	0.62026	37.974	820.14	0.01533	0.02533	21.597	1.6122	61.222	1322.2	48
60	0.55045	44.955	1192.8	0.01224	0.02224	26.533	1.8167	81.669	2166.9	60
120	0.30299	69.700	3334.1	0.00435	0.01435	47.834	3.3003	230.03	11003.	120
180	0.16678	83.321	5330.0	0.00200	0.01200	63.969	5.9958	499.58	31958.	180
240	0.09181	90.819	6878.6	0.00101	0.01101	75.739	10.892	989.25	74925.	240
300	0.05053	94.946	7978.6	0.00053	0.01053	84.032	19.788	1878.8	157885.	300
360	0.02782	97.218	8720.4	0.00029	0.01029	89.699	35.949	3494.9	313496.	360
INF	0.00000	100.00	10000.	0.0000	0.01000	100.00	INF	INF	INF	INF

| | i = 1.5% | | | | i = 1.5% | | | i = 1.5% | | |
| | PRESENT SUM, P | | | UNIFORM SERIES, A | | | FUTURE SUM, F | | | |
n	P/F	P/A	P/G	A/F	A/P	A/G	F/P	F/A	F/G	n
1	0.98522	0.9852	0.0000	1.00000	1.01500	0.0000	1.0150	1.0000	0.0000	1
2	0.97066	1.9558	0.9706	0.49628	0.51128	0.4962	1.0302	2.0150	1.0000	2
3	0.95632	2.9122	2.8833	0.32838	0.34338	0.9900	1.0456	3.0452	3.0150	3
4	0.94218	3.8543	5.7098	0.24444	0.25944	1.4813	1.0613	4.0909	6.0602	4
5	0.92826	4.7826	9.4228	0.19409	0.20909	1.9702	1.0772	5.1522	10.151	5
6	0.91454	5.6971	13.995	0.16053	0.17553	2.4565	1.0934	6.2295	15.303	6
7	0.90103	6.5982	19.401	0.13656	0.15156	2.9404	1.1098	7.3229	21.532	7
8	0.88771	7.4859	25.615	0.11858	0.13358	3.4218	1.1264	8.4328	28.855	8
9	0.87459	8.3605	32.612	0.10461	0.11961	3.9007	1.1433	9.5593	37.288	9
10	0.86167	9.2221	40.367	0.09343	0.10843	4.3772	1.1605	10.702	46.848	10
11	0.84893	10.071	43.856	0.08429	0.09929	4.8511	1.1779	11.863	57.550	11
12	0.83639	10.907	58.057	0.07668	0.09168	5.3226	1.1956	13.041	69.414	12
13	0.82403	11.731	67.945	0.07024	0.08524	5.7916	1.2135	14.236	82.455	13
14	0.81185	12.543	73.499	0.06472	0.07972	6.2582	1.2317	15.450	96.692	14
15	0.79985	13.343	89.697	0.05994	0.07494	6.7223	1.2502	16.682	112.14	15
16	0.78803	14.131	101.51	0.05577	0.07077	7.1839	1.2689	17.932	128.82	16
17	0.77639	14.907	113.94	0.05208	0.06709	7.6430	1.2880	19.201	146.75	17
18	0.76491	15.672	126.94	0.04881	0.06381	8.0997	1.3073	20.489	165.95	18
19	0.75361	16.426	140.50	0.04588	0.06088	8.5539	1.3269	21.796	186.44	19
20	0.74247	17.168	154.61	0.04325	0.05825	9.0056	1.3468	23.123	208.24	20
21	0.73150	17.900	169.24	0.04087	0.05587	9.4549	1.3670	24.470	231.36	21
22	0.72069	18.620	184.38	0.03870	0.05370	9.9018	1.3875	25.837	255.83	22
23	0.71004	19.330	200.00	0.03673	0.05173	10.346	1.4083	27.225	281.67	23
24	0.69954	20.030	216.09	0.03492	0.04992	10.788	1.4295	28.633	308.90	24
25	0.68921	20.719	232.63	0.03326	0.04826	11.227	1.4509	30.063	337.53	25
26	0.67902	21.398	249.60	0.03173	0.04673	11.664	1.4727	31.514	367.59	26
27	0.66899	22.067	267.00	0.03032	0.04532	12.099	1.4948	32.986	399.11	27
28	0.65910	22.726	284.79	0.02900	0.04400	12.531	1.5172	34.481	432.09	28
29	0.64936	23.376	302.97	0.02778	0.04278	12.961	1.5399	35.998	466.58	29
30	0.63976	24.015	321.53	0.02664	0.04164	13.388	1.5630	37.538	502.57	30
32	0.62099	25.267	359.69	0.02458	0.03958	14.235	1.6103	40.688	579.21	32
34	0.60277	26.481	399.16	0.02276	0.03776	15.073	1.6590	43.933	662.20	34
36	0.58509	27.660	439.83	0.02115	0.03615	15.900	1.7091	47.276	751.73	36
48	0.48936	34.042	703.54	0.01437	0.02937	20.666	2.0434	69.565	1437.6	48
60	0.40930	39.380	988.16	0.01039	0.02539	25.093	2.4432	96.214	2414.3	60
120	0.16752	55.498	2359.7	0.00302	0.01802	42.518	5.9693	331.28	14085.	120
180	0.06857	62.095	3316.9	0.00110	0.01610	53.416	14.584	905.62	48375.	180
240	0.02806	64.795	3870.6	0.00043	0.01543	59.736	35.632	2308.8	137924	240
300	0.01189	65.900	4163.6	0.00017	0.01517	63.180	87.058	5737.2	362484.	300
360	0.00470	66.353	4310.7	0.00007	0.01507	64.966	212.70	14113.	916906.	360
INF	0.00000	66.666	4444.4	0.0000	0.01500	66.666	INF	INF	INF	INF

	i = 2% PRESENT SUM, P			i = 2% UNIFORM SERIES, A			i = 2% FUTURE SUM, F			
n	P/F	P/A	P/G	A/F	A/P	A/G	F/P	F/A	F/G	n
1	0.98039	0.9803	0.0000	1.00000	1.02000	0.0000	1.0200	1.0000	0.0000	1
2	0.96117	1.9415	0.9611	0.49505	0.51505	0.4950	1.0404	2.0200	1.0000	2
3	0.94232	2.8838	2.8458	0.32675	0.34675	0.9868	1.0612	3.0604	3.0200	3
4	0.92385	3.8077	5.6173	0.24262	0.26262	1.4752	1.0824	4.1216	6.0804	4
5	0.90573	4.7134	9.2402	0.19216	0.21216	1.9604	1.1040	5.2040	10.202	5
6	0.88797	5.6014	13.680	0.15853	0.17853	2.4422	1.1261	6.3081	15.406	6
7	0.87056	6.4719	18.903	0.13451	0.15451	2.9208	1.1486	7.4342	21.714	7
8	0.85349	7.3254	24.877	0.11651	0.13651	3.3960	1.1716	8.5829	29.148	8
9	0.83676	8.1622	31.572	0.10252	0.12252	3.8680	1.1950	9.7546	37.731	9
10	0.82035	8.9825	38.955	0.09133	0.11133	4.3367	1.2189	10.949	47.486	10
11	0.80426	9.7868	46.997	0.08218	0.10218	4.8021	1.2433	12.168	58.435	11
12	0.78849	10.575	55.671	0.07456	0.09456	5.2642	1.2682	13.412	70.604	12
13	0.77303	11.348	64.947	0.06812	0.08812	5.7230	1.2936	14.680	84.016	13
14	0.75788	12.106	74.799	0.06260	0.08260	6.1786	1.3194	15.973	98.696	14
15	0.74301	12.849	85.202	0.05783	0.07783	6.6309	1.3458	17.293	114.67	15
16	0.72845	13.577	96.128	0.05365	0.07365	7.0799	1.3727	18.639	131.96	16
17	0.71416	14.291	107.55	0.04997	0.06997	7.5256	1.4002	20.012	150.60	17
18	0.70016	14.992	119.45	0.04670	0.06670	7.9681	1.4282	21.412	170.61	18
19	0.68643	15.678	131.81	0.04378	0.06378	8.4073	1.4568	22.840	192.02	19
20	0.67297	16.351	144.60	0.04116	0.06116	8.8432	1.4859	24.297	214.86	20
21	0.65978	17.011	157.79	0.03878	0.05878	9.2759	1.5156	25.783	239.16	21
22	0.64684	17.658	171.37	0.03663	0.05663	9.7054	1.5459	27.299	264.94	22
23	0.63416	18.292	185.33	0.03467	0.05467	10.131	1.5769	28.845	292.24	23
24	0.62172	18.913	199.63	0.03287	0.05287	10.554	1.6084	30.421	321.09	24
25	0.60953	19.523	214.25	0.03122	0.05122	10.974	1.6406	32.030	351.51	25
26	0.59758	20.121	229.19	0.02970	0.04970	11.391	1.6734	33.670	383.54	26
27	0.58586	20.706	244.43	0.02829	0.04829	11.804	1.7068	35.344	417.21	27
28	0.57437	21.281	259.93	0.02699	0.04699	12.214	1.7410	37.051	452.56	28
29	0.56311	21.844	275.70	0.02578	0.04578	12.621	1.7758	38.792	489.61	29
30	0.55207	22.396	291.71	0.02465	0.04465	13.025	1.8113	40.568	528.40	30
32	0.53063	23.468	324.40	0.02261	0.04261	13.823	1.8845	44.227	611.35	32
34	0.51003	24.498	357.88	0.02082	0.04082	14.608	1.9606	48.033	701.69	34
36	0.49022	25.488	392.04	0.01923	0.03923	15.380	2.0398	51.994	799.71	36
48	0.38654	30.673	605.96	0.01250	0.03250	19.755	2.5870	79.353	1567.6	48
60	0.30478	34.760	823.69	0.00877	0.02877	23.696	3.2810	114.05	2702.5	60
120	0.09289	45.355	1710.4	0.00205	0.02205	37.711	10.765	488.25	18412.	120
180	0.02831	48.584	2174.4	0.00058	0.02058	44.755	35.320	1716.0	76802.	180
240	0.00863	49.568	2374.8	0.00017	0.02017	47.911	115.88	5744.4	275222.	240
300	0.00263	49.868	2453.9	0.00005	0.02005	49.208	380.23	18961.	933086.	300
360	0.00080	49.959	2483.5	0.00002	0.02002	49.711	1247.5	62328.	>10**6	360
INF	0.00000	50.000	2500.0	0.0000	0.02000	50.000	INF	INF	INF	INF

| | i = 2.5% | | | i = 2.5% | | | i = 2.5% | | | |
| | PRESENT SUM, P | | | UNIFORM SERIES, A | | | FUTURE SUM, F | | | |
n	P/F	P/A	P/G	A/F	A/P	A/G	F/P	F/A	F/G	n
1	0.97561	0.9756	0.0000	1.00000	1.02500	0.0000	1.0250	1.0000	0.0000	1
2	0.95181	1.9274	0.9518	0.49383	0.51883	0.4938	1.0506	2.0250	1.0000	2
3	0.92860	2.8560	2.8090	0.32514	0.35014	0.9835	1.0768	3.0756	3.0250	3
4	0.90595	3.7619	5.5268	0.24082	0.26582	1.4691	1.1038	4.1525	6.1006	4
5	0.88385	4.6458	9.0622	0.19025	0.21525	1.9506	1.1314	5.2563	10.253	5
6	0.86230	5.5081	13.373	0.15655	0.18155	2.4280	1.1596	6.3877	15.509	6
7	0.84127	6.3493	18.421	0.13250	0.15750	2.9012	1.1886	7.5474	21.897	7
8	0.82075	7.1701	24.166	0.11447	0.13947	3.3704	1.2184	8.7361	29.444	8
9	0.80073	7.9708	30.572	0.10046	0.12546	3.8355	1.2488	9.9545	38.180	9
10	0.78120	8.7520	37.603	0.08926	0.11426	4.2964	1.2800	11.203	48.135	10
11	0.76214	9.5142	45.224	0.08011	0.10511	4.7533	1.3120	12.483	59.338	11
12	0.74356	10.257	53.403	0.07249	0.09749	5.2061	1.3448	13.795	71.822	12
13	0.72542	10.983	62.108	0.06605	0.09105	5.6549	1.3785	15.140	85.617	13
14	0.70773	11.690	71.309	0.06054	0.08554	6.0995	1.4129	16.519	100.75	14
15	0.69047	12.381	80.975	0.05577	0.08077	6.5401	1.4483	17.931	117.27	15
16	0.67362	13.055	91.080	0.05150	0.07660	6.9766	1.4845	19.380	135.20	16
17	0.65720	13.712	101.59	0.04793	0.07293	7.4091	1.5216	20.864	154.58	17
18	0.64117	14.353	112.49	0.04467	0.06967	7.8375	1.5596	22.386	175.45	18
19	0.62553	14.978	123.75	0.04176	0.06676	8.2619	1.5986	23.946	197.84	19
20	0.61027	15.589	135.35	0.03915	0.06415	8.6823	1.6386	25.544	221.73	20
21	0.59539	16.184	147.25	0.03679	0.06179	9.0986	1.6795	27.183	247.33	21
22	0.58086	16.765	159.45	0.03465	0.05965	9.5109	1.7215	28.862	274.51	22
23	0.56670	17.332	171.92	0.03270	0.05770	9.9193	1.7646	30.584	303.37	23
24	0.55288	17.885	184.63	0.03091	0.05591	10.323	1.8087	32.349	333.96	24
25	0.53939	18.424	197.58	0.02928	0.05428	10.724	1.8539	34.157	366.31	25
30	0.47674	20.930	265.12	0.02278	0.04778	12.666	2.0975	43.902	556.10	30
35	0.42137	23.145	335.88	0.01821	0.04321	14.512	2.3732	54.928	797.12	35
36	0.41109	23.556	350.27	0.01745	0.04245	14.869	2.4325	57.301	852.05	36
40	0.37243	25.102	408.22	0.01484	0.03984	16.262	2.6850	67.402	1096.1	40
48	0.30567	27.773	524.03	0.01101	0.03601	18.868	3.2714	90.859	1714.3	48
50	0.29094	28.362	552.60	0.01026	0.03526	19.483	3.4371	97.484	1899.3	50
60	0.22728	30.908	690.86	0.00735	0.03235	22.351	4.3997	135.99	3039.6	60
100	0.08465	36.614	1125.9	0.00231	0.02731	30.752	11.813	432.54	13301.	100
180	0.01174	39.530	1496.6	0.00030	0.02530	37.861	85.171	3366.8	127475.	180
240	0.00267	39.893	1570.1	0.00007	0.02507	39.357	374.73	14949.	588381.	240
300	0.00061	39.975	1591.7	0.00002	0.02502	39.817	1648.7	65910.	>10**6	300
360	0.00014	39.994	1597.7	0.00000	0.02500	39.950	7254.2	290129	>10**6	360
INF	0.00000	40.000	1600.0	0.0000	0.02500	40.000	INF	INF	INF	INF

	i = 3% PRESENT SUM, P			i = 3% UNIFORM SERIES, A			i = 3% FUTURE SUM, F			
n	P/F	P/A	P/G	A/F	A/P	A/G	F/P	F/A	F/G	n
1	0.97087	0.9708	0.0000	1.00000	1.03000	0.0000	1.0300	1.0000	0.0000	1
2	0.94260	1.9134	0.9426	0.49261	0.52261	0.4926	1.0609	2.0300	1.0000	2
3	0.91514	2.8286	2.7728	0.32353	0.35353	0.9803	1.0927	3.0909	3.0300	3
4	0.88849	3.7171	5.4383	0.23903	0.26903	1.4630	1.1255	4.1836	6.1209	4
5	0.86261	4.5797	8.8887	0.18835	0.21835	1.9409	1.1592	5.3091	10.304	5
6	0.83748	5.4171	13.076	0.15460	0.18460	2.4138	1.1940	6.4684	15.613	6
7	0.81309	6.2302	17.954	0.13051	0.16051	2.8818	1.2298	7.6624	22.082	7
8	0.78941	7.0196	23.480	0.11246	0.14246	3.3449	1.2667	8.8923	29.744	8
9	0.76642	7.7861	29.611	0.09843	0.12843	3.8031	1.3047	10.159	38.636	9
10	0.74409	8.5302	36.308	0.08723	0.11723	4.2565	1.3439	11.463	48.796	10
11	0.72242	9.2526	43.533	0.07808	0.10808	4.7049	1.3842	12.807	60.259	11
12	0.70138	9.9540	51.248	0.07046	0.10046	5.1485	1.4257	14.192	73.067	12
13	0.68095	10.635	59.419	0.06403	0.09403	5.5872	1.4685	15.617	87.259	13
14	0.66112	11.296	68.014	0.05853	0.08853	6.0210	1.5125	17.086	102.87	14
15	0.64186	11.937	77.000	0.05377	0.08377	6.4500	1.5579	18.598	119.96	15
16	0.62317	12.561	86.347	0.04961	0.07961	6.8742	1.6047	20.156	138.56	16
17	0.60502	13.166	96.028	0.04595	0.07595	7.2935	1.6528	21.761	158.72	17
18	0.58739	13.753	106.01	0.04271	0.07271	7.7081	1.7024	23.414	180.48	18
19	0.57029	14.323	116.27	0.03981	0.06981	8.1178	1.7535	25.116	203.89	19
20	0.55368	14.877	126.79	0.03722	0.06722	8.5228	1.8061	26.870	229.01	20
21	0.53755	15.415	137.55	0.03487	0.06487	8.9230	1.8602	28.676	255.88	21
22	0.52189	15.936	148.50	0.03275	0.06275	9.3185	1.9161	30.536	284.55	22
23	0.50669	16.443	159.65	0.03081	0.06081	9.7093	1.9735	32.452	315.09	23
24	0.49193	16.935	170.97	0.02905	0.05905	10.095	2.0327	34.426	347.54	24
25	0.47761	17.413	182.43	0.02743	0.05743	10.476	2.0937	36.459	381.97	25
26	0.46369	17.876	194.02	0.02594	0.05594	10.853	2.1565	38.553	418.43	26
28	0.43708	18.764	217.53	0.02329	0.05329	11.593	2.2879	42.930	497.69	28
30	0.41199	19.600	241.36	0.02102	0.05102	12.314	2.4272	47.575	585.84	30
35	0.35538	21.487	301.62	0.01654	0.04654	14.037	2.8138	60.462	848.73	35
36	0.34503	21.832	313.70	0.01580	0.04580	14.368	2.8982	63.275	909.19	36
40	0.30656	23.114	351.75	0.01326	0.04326	15.650	3.2620	75.401	1180.0	40
45	0.26444	24.518	420.63	0.01079	0.04079	17.155	3.7816	92.719	1590.6	45
48	0.24200	25.266	455.02	0.00958	0.03958	18.008	4.1322	104.40	1880.2	48
50	0.22811	25.729	477.48	0.00887	0.03887	18.557	4.3839	112.79	2093.2	50
60	0.16973	27.675	583.05	0.00613	0.03613	21.067	5.8916	163.05	3435.1	60
70	0.12630	29.123	676.08	0.00434	0.03434	23.214	7.9178	230.59	5353.1	70
80	0.09398	30.200	756.08	0.00311	0.03311	25.035	10.640	321.36	8045.4	80
90	0.06993	31.002	823.63	0.00226	0.03226	26.566	14.300	443.34	11778.	90
100	0.05203	31.598	879.85	0.00165	0.03165	27.844	19.218	607.28	16909.	100
120	0.02881	32.373	963.86	0.00089	0.03089	29.773	34.711	1123.7	33456.	120
180	0.00489	33.170	1076.3	0.00015	0.03015	32.448	204.50	6783.4	220115.	180
240	0.00083	33.305	1103.5	0.00002	0.03002	33.134	1204.8	40128.	>10**6	240
INF	0.00000	33.333	1111.1	0.0000	0.03000	33.333	INF	INF	INF	INF

	i = 4%			i = 4%			i = 4%			
	PRESENT SUM, P			UNIFORM SERIES, A			FUTURE SUM, F			
n	P/F	P/A	P/G	A/F	A/P	A/G	F/P	F/A	F/G	n
1	0.96154	0.9615	0.0000	1.00000	1.04000	0.0000	1.0400	1.0000	0.0000	1
2	0.92456	1.8860	0.9245	0.49020	0.53020	0.4902	1.0816	2.0400	1.0000	2
3	0.88900	2.7750	2.7025	0.32035	0.36035	0.9738	1.1248	3.1216	3.0400	3
4	0.85480	3.6299	5.2669	0.23549	0.27549	1.4510	1.1698	4.2464	6.1616	4
5	0.82193	4.4518	8.5546	0.18463	0.22463	1.9216	1.2166	5.4163	10.408	5
6	0.79031	5.2421	12.506	0.15076	0.19076	2.3857	1.2653	6.6329	15.824	6
7	0.75992	6.0020	17.065	0.12661	0.16661	2.8433	1.3159	7.8982	22.457	7
8	0.73069	6.7327	22.180	0.10853	0.14853	3.2944	1.3685	9.2142	30.355	8
9	0.70259	7.4353	27.801	0.09449	0.13449	3.7390	1.4233	10.582	39.569	9
10	0.67556	8.1109	33.881	0.08329	0.12329	4.1772	1.4802	12.006	50.152	10
11	0.64958	8.7604	40.377	0.07415	0.11415	4.6090	1.5394	13.486	62.158	11
12	0.62460	9.3850	47.247	0.06655	0.10655	5.0343	1.6010	15.025	75.645	12
13	0.60057	9.9856	54.454	0.06014	0.10014	5.4532	1.6650	16.626	90.670	13
14	0.57748	10.563	61.961	0.05467	0.09467	5.8658	1.7316	18.291	107.29	14
15	0.55526	11.118	69.735	0.04994	0.08994	6.2720	1.8009	20.023	125.59	15
16	0.53391	11.652	77.744	0.04582	0.08582	6.6720	1.8729	21.824	145.61	16
17	0.51337	12.165	85.958	0.04220	0.08220	7.0656	1.9479	23.697	167.43	17
18	0.49363	12.659	94.349	0.03899	0.07899	7.4530	2.0258	25.645	191.13	18
19	0.47464	13.134	102.89	0.03614	0.07614	7.8341	2.1068	27.671	216.78	19
20	0.45639	13.590	111.56	0.03358	0.07358	8.2091	2.1911	29.778	244.45	20
21	0.43883	14.029	120.34	0.03128	0.07128	8.5779	2.2787	31.969	274.23	21
22	0.42196	14.451	129.20	0.02920	0.06920	8.9406	2.3699	34.248	306.19	22
23	0.40573	14.856	138.12	0.02731	0.06731	9.2972	2.4647	36.617	340.44	23
24	0.39012	15.247	147.10	0.02559	0.06559	9.6479	2.5633	39.082	377.06	24
25	0.37512	15.622	156.10	0.02401	0.06401	9.9925	2.6658	41.645	416.14	25
26	0.36069	15.982	165.12	0.02257	0.06257	10.331	2.7724	44.311	457.79	26
28	0.33348	16.663	183.14	0.02001	0.06001	10.990	2.9987	49.967	549.19	28
30	0.30832	17.292	201.06	0.01783	0.05783	11.627	3.2434	56.084	652.12	30
35	0.25342	18.664	244.87	0.01358	0.05358	13.119	3.9460	73.652	966.30	35
36	0.24367	18.908	253.40	0.01289	0.05289	13.401	4.1039	77.598	1039.9	36
40	0.20829	19.792	286.53	0.01052	0.05052	14.476	4.8010	95.025	1375.6	40
45	0.17120	20.720	325.40	0.00826	0.04826	15.704	5.8411	121.02	1900.7	45
48	0.15219	21.195	347.24	0.00718	0.04718	16.383	6.5705	139.26	2281.5	48
50	0.14071	21.482	361.16	0.00655	0.04655	16.812	7.1066	152.66	2566.6	50
55	0.11566	22.108	393.68	0.00523	0.04523	17.807	8.6463	191.15	3403.9	55
60	0.09506	22.623	422.99	0.00420	0.04420	18.697	10.519	237.99	4449.7	60
70	0.06422	23.394	472.47	0.00275	0.04275	20.196	15.571	364.29	7357.2	70
80	0.04338	23.915	511.11	0.00181	0.04181	21.371	23.049	551.24	11781.	80
90	0.02931	24.267	540.73	0.00121	0.04121	22.283	34.119	827.98	18449.	90
100	0.01980	24.505	563.12	0.00081	0.04081	22.980	50.504	1237.6	28440.	100
120	0.00904	24.774	592.24	0.00036	0.04036	23.905	110.66	2741.5	65539.	120
180	0.00086	24.978	620.59	0.00003	0.04003	24.845	1164.1	29078.	722456.	180
INF	0.00000	25.000	625.00	0.0000	0.04000	25.000	INF	INF	INF	INF

$i = 5\%$

	PRESENT SUM, P			UNIFORM SERIES, A			FUTURE SUM, F			
n	P/F	P/A	P/G	A/F	A/P	A/G	F/P	F/A	F/G	n
1	0.95238	0.9523	0.0000	1.00000	1.05000	0.0000	1.0500	1.0000	0.0000	1
2	0.90703	1.8594	0.9070	0.48780	0.53780	0.4878	1.1025	2.0500	1.0000	2
3	0.86384	2.7232	2.6347	0.31721	0.36721	0.9674	1.1576	3.1525	3.0500	3
4	0.82270	3.5459	5.1028	0.23201	0.28201	1.4390	1.2155	4.3101	6.2025	4
5	0.78353	4.3294	8.2369	0.18097	0.23097	1.9025	1.2762	5.5256	10.512	5
6	0.74622	5.0756	11.968	0.14702	0.19702	2.3579	1.3401	6.8019	16.038	6
7	0.71068	5.7863	16.232	0.12282	0.17282	2.8052	1.4071	8.1420	22.840	7
8	0.67684	6.4632	20.270	0.10472	0.15472	3.2245	1.4774	9.5491	30.982	8
9	0.64461	7.1078	26.126	0.09069	0.14069	3.6757	1.5513	11.026	40.531	9
10	0.61391	7.7217	31.652	0.07950	0.12950	4.0990	1.6288	12.577	51.557	10
11	0.58468	8.3064	37.498	0.07039	0.12039	4.5144	1.7103	14.206	64.135	11
12	0.55684	8.8632	43.624	0.06283	0.11283	4.9219	1.7958	15.917	78.342	12
13	0.53032	9.3935	49.987	0.05646	0.10646	5.3215	1.8856	17.713	94.259	13
14	0.50507	9.8986	56.553	0.05102	0.10102	5.7132	1.9799	19.598	111.97	14
15	0.48102	10.379	63.288	0.04634	0.09634	6.0973	2.0789	21.578	131.57	15
16	0.45811	10.837	70.159	0.04227	0.09227	6.4736	2.1828	23.657	153.15	16
17	0.43630	11.274	77.140	0.03870	0.08870	6.8422	2.2920	25.840	176.80	17
18	0.41552	11.689	84.204	0.03555	0.08555	7.2033	2.4066	28.132	202.64	18
19	0.39573	12.085	91.327	0.03275	0.08275	7.5569	2.5269	30.539	230.78	19
20	0.37689	12.462	98.488	0.03024	0.08024	7.9029	2.6533	33.066	261.31	20
21	0.35894	12.821	105.66	0.02800	0.07800	8.2416	2.7859	35.719	294.38	21
22	0.34185	13.163	112.84	0.02597	0.07597	8.5729	2.9252	38.505	330.10	22
23	0.32557	13.488	120.00	0.02414	0.07414	8.8970	3.0715	41.430	368.61	23
24	0.31007	13.798	127.14	0.02247	0.07247	9.2139	3.2251	44.502	410.04	24
25	0.29530	14.093	134.22	0.02095	0.07095	9.5237	3.3863	47.727	454.54	25
26	0.28124	14.375	141.25	0.01956	0.06956	9.8265	3.5556	51.113	502.26	26
28	0.25509	14.898	155.11	0.01712	0.06712	10.411	3.9201	58.402	608.05	28
30	0.23138	15.372	168.62	0.01505	0.06505	10.969	4.3219	66.438	728.77	30
35	0.18129	16.374	200.58	0.01107	0.06107	12.249	5.5160	90.320	1106.4	35
36	0.17266	16.546	206.62	0.01043	0.06043	12.487	5.7918	95.836	1196.7	36
40	0.14205	17.159	229.54	0.00828	0.05828	13.377	7.0399	120.80	1616.0	40
45	0.11130	17.774	255.31	0.00626	0.05626	14.364	8.9850	159.70	2294.0	45
48	0.09614	18.077	269.24	0.00532	0.05532	14.894	10.401	188.02	2800.5	48
50	0.08720	18.255	279.51	0.00478	0.05478	15.223	11.467	209.34	3186.9	50
55	0.06833	18.633	297.51	0.00367	0.05367	15.966	14.635	272.71	4354.2	55
60	0.05354	18.929	314.34	0.00283	0.05283	16.606	18.679	353.58	5871.6	60
70	0.03287	19.342	340.84	0.00170	0.05170	17.621	30.426	588.52	10370.	70
80	0.02018	19.596	359.64	0.00103	0.05103	18.352	49.561	971.22	17824.	80
90	0.01239	19.752	372.74	0.00063	0.05063	18.871	80.730	1594.6	30092.	90
100	0.00760	19.847	381.74	0.00038	0.05038	19.233	131.50	2610.0	50200.	100
120	0.00287	19.942	391.97	0.00014	0.05014	19.655	348.91	6958.2	136765.	120
INF	0.00000	20.000	400.00	0.0000	0.05000	20.000	INF	INF	INF	INF

	i = 6% PRESENT SUM, P			i = 6% UNIFORM SERIES, A			i = 6% FUTURE SUM, F			
n	P/F	P/A	P/G	A/F	A/P	A/G	F/P	F/A	F/G	n
1	0.94340	0.9434	0.0000	1.00000	1.06000	0.0000	1.0600	1.0000	0.0000	1
2	0.89000	1.8334	0.8900	0.48544	0.54544	0.4854	1.1236	2.0600	1.0000	2
3	0.83962	2.6730	2.5692	0.31411	0.37411	0.9611	1.1910	3.1836	3.0600	3
4	0.79209	3.4651	4.9455	0.22859	0.28859	1.4272	1.2624	4.3746	6.2436	4
5	0.74726	4.2123	7.9345	0.17740	0.23740	1.8836	1.3382	5.6370	10.618	5
6	0.70496	4.9173	11.459	0.14336	0.20336	2.3304	1.4185	6.9753	16.255	6
7	0.66506	5.5823	15.449	0.11914	0.17914	2.7675	1.5036	8.3938	23.230	7
8	0.62741	6.2097	19.841	0.10104	0.16104	3.1952	1.5938	9.8974	31.624	8
9	0.59190	6.8016	24.576	0.08702	0.14702	3.6133	1.6894	11.491	41.521	9
10	0.55839	7.3600	29.602	0.07587	0.13587	4.0220	1.7908	13.180	53.013	10
11	0.52679	7.8868	34.870	0.06679	0.12679	4.4212	1.8983	14.971	66.194	11
12	0.49697	8.3838	40.336	0.05928	0.11928	4.8112	2.0122	16.869	81.165	12
13	0.46884	8.8526	45.962	0.05296	0.11296	5.1919	2.1329	18.882	98.035	13
14	0.44230	9.2949	51.712	0.04758	0.10758	5.5635	2.2609	21.015	116.91	14
15	0.41727	9.7122	57.554	0.04296	0.10296	5.9259	2.3965	23.276	137.93	15
16	0.39365	10.105	63.459	0.03895	0.09895	6.2794	2.5403	25.672	161.20	16
17	0.37136	10.477	69.401	0.03544	0.09544	6.6239	2.6927	28.212	186.88	17
18	0.35034	10.857	75.356	0.03236	0.09236	6.9597	2.8543	30.905	215.09	18
19	0.33051	11.158	81.306	0.02962	0.08962	7.2867	3.0256	33.760	246.00	19
20	0.31180	11.469	87.230	0.02718	0.08718	7.6051	3.2071	36.785	279.76	20
21	0.29416	11.764	93.113	0.02500	0.08500	7.9150	3.3995	39.992	316.54	21
22	0.27751	12.303	98.941	0.02305	0.08305	8.2166	3.6035	43.392	356.53	22
23	0.26180	12.303	104.70	0.02128	0.08128	8.5099	3.8197	46.995	399.93	23
24	0.24698	12.550	110.38	0.01968	0.07968	8.7950	4.0489	50.815	446.92	24
25	0.23300	12.783	115.97	0.01823	0.07823	9.0722	4.2918	54.864	497.74	25
26	0.21981	13.003	121.46	0.01690	0.07690	9.3414	4.5493	59.156	552.60	26
27	0.20737	13.206	126.86	0.01570	0.07570	9.6029	4.8223	63.705	611.76	27
28	0.19563	13.406	132.14	0.01459	0.07459	9.8568	5.1116	68.528	675.46	28
29	0.18456	13.590	137.31	0.01358	0.07358	10.103	5.4183	73.639	743.99	29
30	0.17411	13.764	142.35	0.01265	0.07265	10.342	5.7434	79.058	817.63	30
35	0.13011	14.498	165.74	0.00897	0.06897	11.431	7.6860	111.43	1273.9	35
40	0.09722	15.046	185.95	0.00646	0.06646	12.359	10.285	154.76	1912.7	40
45	0.07265	15.456	203.11	0.00470	0.06470	13.141	13.764	212.74	2795.7	45
50	0.05429	15.761	217.45	0.00344	0.06344	13.796	18.420	290.33	4005.6	50
55	0.04057	15.990	229.32	0.00254	0.06254	14.341	24.650	394.17	5652.8	55
60	0.03031	16.161	239.04	0.00188	0.06188	14.790	32.987	533.12	7885.4	60
65	0.02265	16.289	246.94	0.00139	0.06139	15.160	44.145	719.08	10901.	65
70	0.01693	16.384	252.32	0.00103	0.06103	15.461	59.075	967.93	14965.	70
80	0.00945	16.509	262.54	0.00057	0.06057	15.903	105.79	1746.6	27776.	80
90	0.00528	16.578	268.39	0.00032	0.06032	16.189	189.46	3141.0	50851.	90
100	0.00295	16.617	272.04	0.00018	0.06018	16.371	339.30	5638.3	92306.	100
120	0.00092	16.651	275.68	0.00006	0.06006	16.556	1088.1	18119.	299997.	120
INF	0.00000	16.666	277.77	0.0000	0.06000	16.666	INF	INF	INF	INF

	i = 7%			i = 7%			i = 7%			
	PRESENT SUM, P			UNIFORM SERIES, A			FUTURE SUM, F			
n	P/F	P/A	P/G	A/F	A/P	A/G	F/P	F/A	F/G	n
1	0.93458	0.9345	0.0000	1.00000	1.07000	0.0000	1.0700	1.0000	0.0000	1
2	0.87344	1.8080	0.8734	0.48309	0.55309	0.4830	1.1449	2.0700	1.0000	2
3	0.81630	2.6243	2.5060	0.31105	0.38105	0.9549	1.2250	3.2149	3.0700	3
4	0.76290	3.3872	4.7947	0.22523	0.29523	1.4155	1.3108	4.4399	6.2849	4
5	0.71299	4.1002	7.6466	0.17389	0.24389	1.8649	1.4025	5.7507	10.724	5
6	0.66634	4.7665	10.978	0.13980	0.20980	2.3032	1.5007	7.1532	16.475	6
7	0.62275	5.3892	14.714	0.11555	0.18555	2.7303	1.6057	8.6540	23.628	7
8	0.58201	5.9713	18.788	0.09747	0.16747	3.1465	1.7181	10.258	32.282	8
9	0.54393	6.5152	23.140	0.08349	0.15349	3.5517	1.8384	11.978	42.542	9
10	0.50835	7.0235	27.715	0.07238	0.14238	3.9460	1.9671	13.816	54.520	10
11	0.47509	7.4986	32.466	0.06336	0.13336	4.3296	2.1048	15.783	68.337	11
12	0.44401	7.9426	37.350	0.05590	0.12590	4.7025	2.2521	17.888	84.120	12
13	0.41496	8.3576	42.330	0.04965	0.11965	5.0648	2.4098	20.140	102.00	13
14	0.38782	8.7454	47.371	0.04434	0.11434	5.4167	2.5785	22.550	122.15	14
15	0.36245	9.1079	52.446	0.03979	0.10979	5.7582	2.7590	25.129	144.70	15
16	0.33873	9.4466	57.527	0.03586	0.10586	6.0896	2.9521	27.888	169.82	16
17	0.31657	9.7632	62.592	0.03243	0.10243	6.4110	3.1588	30.840	197.71	17
18	0.29586	10.059	67.621	0.02941	0.09941	6.7224	3.3799	33.999	228.55	18
19	0.27651	10.335	72.599	0.02675	0.09675	7.0241	3.6165	37.379	262.55	19
20	0.25842	10.594	77.509	0.02439	0.09439	7.3163	3.8696	40.995	299.93	20
21	0.24151	10.835	82.339	0.02229	0.09229	7.5990	4.1405	44.865	340.93	21
22	0.22571	11.061	87.079	0.02041	0.09041	7.8724	4.4304	49.005	385.79	22
23	0.21095	11.272	91.720	0.01871	0.08871	8.1368	4.7405	53.436	434.80	23
24	0.19715	11.469	96.254	0.01719	0.08719	8.3923	5.0723	58.176	488.23	24
25	0.18425	11.653	100.67	0.01581	0.08581	8.6391	5.4274	63.249	546.41	25
26	0.17220	11.825	104.98	0.01456	0.08456	8.8773	5.8073	68.676	609.66	26
27	0.16093	11.986	109.16	0.01343	0.08343	9.1072	6.2138	74.483	678.34	27
28	0.15040	12.137	113.22	0.01239	0.08239	9.3289	6.6488	80.697	752.82	28
29	0.14056	12.277	117.16	0.01145	0.08145	9.5427	7.1142	87.346	833.52	29
30	0.13137	12.409	120.97	0.01059	0.08059	9.7486	7.6122	94.460	920.86	30
35	0.09366	12.947	138.13	0.00723	0.07723	10.668	10.676	138.23	1474.8	35
40	0.06678	13.331	152.29	0.00501	0.07501	11.423	14.974	199.63	2280.5	40
45	0.04761	13.605	163.75	0.00350	0.07350	12.036	21.002	285.74	3439.2	45
50	0.03395	13.800	172.90	0.00246	0.07246	12.528	29.457	406.52	5093.2	50
55	0.02420	13.939	180.12	0.00174	0.07174	12.921	41.315	575.92	7441.8	55
60	0.01726	14.039	185.76	0.00123	0.07123	13.232	57.946	813.52	10764.	60
65	0.01230	14.109	190.14	0.00087	0.07087	13.476	81.272	1146.7	15453.	65
70	0.00877	14.160	193.51	0.00062	0.07062	13.666	113.98	1614.1	22059.	70
80	0.00446	14.222	198.07	0.00031	0.07031	13.927	224.23	3189.0	44415.	80
90	0.00227	14.253	200.70	0.00016	0.07016	14.081	441.10	6287.1	88531.	90
100	0.00115	14.269	202.20	0.00008	0.07008	14.170	867.71	12381.	175452.	100
120	0.00030	14.281	203.51	0.00002	0.07002	14.250	3357.7	47954.	683345.	120
INF	0.00000	14.285	204.08	0.0000	0.07000	14.285	INF	INF	INF	INF

$i = 8\%$

	PRESENT SUM, P			UNIFORM SERIES, A			FUTURE SUM, F			
n	P/F	P/A	P/G	A/F	A/P	A/G	F/P	F/A	F/G	n
1	0.92593	0.9259	0.0000	1.00000	1.08000	0.0000	1.0800	1.0000	0.0000	1
2	0.85734	1.7832	0.8573	0.48077	0.56077	0.4807	1.1664	2.0800	1.0000	2
3	0.79383	2.5771	2.4450	0.30803	0.38803	0.9487	1.2597	3.2464	3.0800	3
4	0.73503	3.3121	4.6500	0.22192	0.30192	1.4039	1.3604	4.5061	6.3264	4
5	0.68058	3.9927	7.3724	0.17046	0.25046	1.8464	1.4693	5.8666	10.832	5
6	0.63017	4.6228	10.523	0.13632	0.21632	2.2763	1.5868	7.3359	16.699	6
7	0.58349	5.2063	14.024	0.11207	0.19207	2.6936	1.7138	8.9228	24.035	7
8	0.54027	5.7466	17.806	0.09401	0.17401	3.0985	1.8509	10.636	32.957	8
9	0.50025	6.2468	21.808	0.08008	0.16008	3.4910	1.9990	12.487	43.594	9
10	0.46319	6.7100	25.976	0.06903	0.14903	3.8713	2.1589	14.486	56.082	10
11	0.42888	7.1389	30.265	0.06008	0.14008	4.2395	2.3316	16.645	70.568	11
12	0.39711	7.5360	34.633	0.05270	0.13270	4.5957	2.5181	18.977	87.214	12
13	0.36770	7.9037	39.046	0.04652	0.12652	4.9402	2.7196	21.495	106.19	13
14	0.34046	8.2442	43.472	0.04130	0.12130	5.2730	2.9371	24.214	127.68	14
15	0.31524	8.5594	47.885	0.03683	0.11683	5.5944	3.1721	27.152	151.90	15
16	0.29189	8.8513	52.264	0.03298	0.11298	5.9046	3.4259	30.324	179.05	16
17	0.27025	9.1216	56.588	0.02963	0.10963	6.2037	3.7000	33.750	209.37	17
18	0.25025	9.3718	60.842	0.02670	0.10670	6.4920	3.9960	37.450	243.12	18
19	0.23171	9.6036	65.013	0.02413	0.10413	6.7696	4.3157	41.446	280.57	19
20	0.21455	9.8181	69.089	0.02185	0.10185	7.0369	4.6609	45.762	322.02	20
21	0.19866	10.016	73.062	0.01983	0.09983	7.2940	5.0338	50.422	367.78	21
22	0.18394	10.200	76.925	0.01803	0.09803	7.5411	5.4365	55.456	418.20	22
23	0.17032	10.371	80.672	0.01642	0.09642	7.7786	5.8714	60.893	473.66	23
24	0.15770	10.528	84.299	0.01498	0.09498	8.0066	6.3411	66.764	534.55	24
25	0.14602	10.674	87.804	0.01368	0.09368	8.2253	6.8484	73.105	601.32	25
26	0.13520	10.810	91.184	0.01251	0.09251	8.4351	7.3963	79.954	674.43	26
27	0.12519	10.935	94.454	0.01145	0.09145	8.6362	7.9880	87.350	754.38	27
28	0.11591	11.051	97.568	0.01049	0.09049	8.8288	8.6271	95.338	841.73	28
29	0.10733	11.158	100.57	0.00962	0.08962	9.0132	9.3172	103.96	937.07	29
30	0.09938	11.257	103.45	0.00883	0.08883	9.1897	10.062	113.28	1041.0	30
35	0.06763	11.654	116.09	0.00580	0.08580	9.9610	14.785	172.31	1716.4	35
40	0.04603	11.924	126.04	0.00386	0.08386	10.569	21.724	259.05	2738.2	40
45	0.03133	12.108	133.73	0.00259	0.08259	11.044	31.920	386.50	4268.8	45
50	0.02132	12.233	139.59	0.00174	0.08174	11.410	46.901	573.77	6547.1	50
55	0.01451	12.318	144.00	0.00118	0.08118	11.690	68.913	848.92	9924.0	55
60	0.00988	12.376	147.30	0.00080	0.08080	11.901	101.25	1253.2	14915.	60
65	0.00672	12.416	149.73	0.00054	0.08054	12.060	148.60	1847.2	22278.	65
70	0.00457	12.442	151.53	0.00037	0.08037	12.178	218.60	2720.0	33126.	70
80	0.00212	12.473	153.80	0.00017	0.08017	12.330	471.95	5886.9	72586.	80
90	0.00098	12.487	154.99	0.00008	0.08008	12.411	1018.9	12723.	157924.	90
100	0.00045	12.494	155.61	0.00004	0.08004	12.454	2199.7	27484.	342306.	100
120	0.00010	12.498	156.08	0.00001	0.08001	12.478	10253.	128150.	>10**6	120
INF	0.00000	12.500	156.25	0.0000	0.08000	12.500	INF	INF	INF	INF

	$i=9\%$				$i=9\%$			$i=9\%$		
	PRESENT SUM, P				UNIFORM SERIES, A			FUTURE SUM, F		
n	P/F	P/A	P/G	A/F	A/P	A/G	F/P	F/A	F/G	n
1	0.91743	0.9174	0.0000	1.00000	1.09000	0.0000	1.0900	1.0000	0.0000	1
2	0.84168	1.7591	0.8416	0.47847	0.56847	0.4784	1.1881	2.0900	1.0000	2
3	0.77218	2.5312	2.3860	0.30505	0.39505	0.9426	1.2950	3.2781	3.0900	3
4	0.70843	3.2397	4.5113	0.21867	0.30867	1.3925	1.4115	4.5731	6.3681	4
5	0.64993	3.8896	7.1110	0.16709	0.25709	1.8282	1.5386	5.9847	10.941	5
6	0.59627	4.4859	10.092	0.13292	0.22292	2.2497	1.6771	7.5233	16.925	6
7	0.54703	5.0329	13.374	0.10869	0.19869	2.6574	1.8280	9.2004	24.449	7
8	0.50187	5.5348	16.887	0.09067	0.18067	3.0511	1.9925	11.028	33.649	8
9	0.46043	5.9952	20.571	0.07680	0.16680	3.4312	2.1718	13.021	44.678	9
10	0.42241	6.4176	24.372	0.06582	0.15582	3.7977	2.3673	15.192	57.699	10
11	0.38753	6.8051	28.248	0.05695	0.14695	4.1509	2.5804	17.560	72.892	11
12	0.35553	7.1607	32.159	0.04965	0.13965	4.4910	2.8126	20.140	90.452	12
13	0.32618	7.4869	36.073	0.04357	0.13357	4.8181	3.0658	22.953	110.59	13
14	0.29925	7.7861	39.963	0.03843	0.12843	5.1326	3.3417	26.019	133.54	14
15	0.27454	8.0606	43.806	0.03406	0.12406	5.4346	3.6424	29.360	159.56	15
16	0.25187	8.3125	47.584	0.03030	0.12030	5.7244	3.9703	33.003	188.92	16
17	0.23107	8.5436	51.282	0.02705	0.11705	6.0023	4.3276	36.973	221.93	17
18	0.21199	8.7556	54.886	0.02421	0.11421	6.2686	4.7171	41.301	258.90	18
19	0.19449	8.9501	58.386	0.02173	0.11173	6.5235	5.1416	46.018	300.20	19
20	0.17843	9.1285	61.777	0.01955	0.10955	6.7674	5.6044	51.160	346.22	20
21	0.16370	9.2922	65.050	0.01762	0.10762	7.0005	6.1088	56.764	397.38	21
22	0.15018	9.4424	68.204	0.01590	0.10590	7.2232	6.6586	62.873	454.14	22
23	0.13778	9.5802	71.235	0.01438	0.10438	7.4357	7.2578	69.531	517.02	23
24	0.12640	9.7066	74.143	0.01302	0.10302	7.6384	7.9110	76.789	586.55	24
25	0.11597	9.8225	76.926	0.01181	0.10181	7.8316	8.6230	84.700	663.34	25
26	0.10639	9.9289	79.586	0.01072	0.10072	8.0155	9.3991	93.324	748.04	26
27	0.09761	10.026	82.124	0.00973	0.09973	8.1906	10.245	102.72	841.36	27
28	0.08955	10.116	84.541	0.00885	0.09885	8.3571	11.167	112.96	944.09	28
29	0.08215	10.198	86.842	0.00806	0.09806	8.5153	12.172	124.13	1057.0	29
30	0.07537	10.273	89.028	0.00734	0.09734	8.6656	13.267	136.30	1181.1	30
35	0.04899	10.566	98.359	0.00464	0.09464	9.3082	20.414	215.71	2007.9	35
40	0.03184	10.757	105.37	0.00296	0.09296	9.7957	31.409	337.88	3309.8	40
45	0.02069	10.881	111.55	0.00190	0.09190	10.160	48.327	525.85	5342.8	45
50	0.01345	10.961	114.32	0.00123	0.09123	10.429	74.357	815.08	8500.9	50
55	0.00874	11.014	117.03	0.00079	0.09079	10.626	114.40	1260.0	13389.	55
60	0.00568	11.048	118.96	0.00051	0.09051	10.768	176.03	1944.7	20942.	60
65	0.00369	11.070	120.33	0.00033	0.09033	10.870	270.84	2998.2	32592.	65
70	0.00240	11.084	121.29	0.00022	0.09022	10.942	416.73	4619.2	50546.	70
80	0.00101	11.099	122.43	0.00009	0.09009	11.029	986.55	10950.	120784.	80
90	0.00043	11.106	122.97	0.00004	0.09004	11.072	2335.5	25939.	287213.	90
100	0.00018	11.109	123.23	0.00002	0.09002	11.093	5529.0	61422.	681363.	100
120	0.00003	11.110	123.41	0.00000	0.09000	11.107	30987.	344289.	>10*+6	120
INF	0.00000	11.111	123.45	0.0000	0.09000	11.111	INF	INF	INF	INF

| | i=10% | | | i=10% | | | i=10% | | | |
| | PRESENT SUM, P | | | UNIFORM SERIES, A | | | FUTURE SUM, F | | | |
n	P/F	P/A	P/G	A/F	A/P	A/G	F/P	F/A	F/G	n
1	0.90909	0.9090	0.0000	1.00000	1.10000	0.0000	1.1000	1.0000	0.0000	1
2	0.82645	1.7355	0.8264	0.47619	0.57619	0.4761	1.2100	2.1000	1.0000	2
3	0.75131	2.4868	2.3290	0.30211	0.40211	0.9365	1.3310	3.3100	3.1000	3
4	0.68301	3.1698	4.3781	0.21547	0.31547	1.3811	1.4641	4.6410	6.4100	4
5	0.62092	3.7907	6.8618	0.16380	0.26380	1.8101	1.6105	6.1051	11.051	5
6	0.56447	4.3552	9.6841	0.12961	0.22961	2.2235	1.7715	7.7156	17.156	6
7	0.51316	4.8684	12.763	0.10541	0.20541	2.6216	1.9487	9.4871	24.871	7
8	0.46651	5.3349	16.028	0.08744	0.18744	3.0044	2.1435	11.435	34.358	8
9	0.42410	5.7590	19.421	0.07364	0.17364	3.3723	2.3579	13.579	45.794	9
10	0.38554	6.1445	22.891	0.06275	0.16275	3.7254	2.5937	15.937	59.374	10
11	0.35049	6.4950	26.396	0.05396	0.15396	4.0640	2.8531	18.531	75.311	11
12	0.31863	6.8136	29.901	0.04676	0.14676	4.3884	3.1384	21.384	93.842	12
13	0.28966	7.1033	33.377	0.04078	0.14078	4.6987	3.4522	24.522	115.22	13
14	0.26333	7.3666	36.800	0.03575	0.13575	4.9955	3.7975	27.975	139.25	14
15	0.23939	7.6060	40.152	0.03147	0.13147	5.2789	4.1772	31.772	167.72	15
16	0.21763	7.8237	43.416	0.02782	0.12782	5.5493	4.5949	35.949	199.49	16
17	0.19784	8.0215	46.581	0.02466	0.12466	5.8071	5.0544	40.544	235.44	17
18	0.17986	8.2014	49.639	0.02193	0.12193	6.0525	5.5599	45.599	275.99	18
19	0.16351	8.3649	52.582	0.01955	0.11955	6.2861	6.1159	51.159	321.59	19
20	0.14864	8.5135	55.406	0.01746	0.11746	6.5080	6.7275	57.275	372.75	20
21	0.13513	8.6486	58.109	0.01562	0.11562	6.7188	7.4002	64.002	430.02	21
22	0.12285	8.7715	60.689	0.01401	0.11401	6.9188	8.1402	71.402	494.02	22
23	0.11168	8.8832	63.146	0.01257	0.11257	7.1084	8.9543	79.543	565.43	23
24	0.10153	8.9847	65.481	0.01130	0.11130	7.2880	9.8497	88.497	644.97	24
25	0.09230	9.0770	67.696	0.01017	0.11017	7.4579	10.834	98.347	733.47	25
26	0.08391	9.1609	69.794	0.00916	0.10916	7.6186	11.918	109.18	831.81	26
27	0.07628	9.2372	71.777	0.00826	0.10826	7.7704	13.110	121.10	940.99	27
28	0.06934	9.3065	73.649	0.00745	0.10745	7.9137	14.421	134.21	1062.1	28
29	0.06304	9.3696	75.414	0.00673	0.10673	8.0488	15.863	148.63	1196.3	29
30	0.05731	9.4269	77.076	0.00608	0.10608	8.1762	17.449	164.49	1344.9	30
35	0.03558	9.6441	83.987	0.00369	0.10369	8.7086	28.102	271.02	2360.2	35
40	0.02209	9.7790	88.952	0.00226	0.10226	9.0962	45.259	442.59	4025.9	40
45	0.01372	9.8628	92.454	0.00139	0.10139	9.3740	72.890	718.90	6739.0	45
50	0.00852	9.9148	94.888	0.00086	0.10086	9.5704	117.39	1163.9	11139.	50
55	0.00529	9.9471	96.561	0.00053	0.10053	9.7075	189.05	1880.5	18255.	55
60	0.00328	9.9671	97.701	0.00033	0.10033	9.8022	304.48	3034.8	29748.	60
65	0.00204	9.9796	98.471	0.00020	0.10020	9.8671	490.37	4893.7	48287.	65
70	0.00127	9.9873	98.987	0.00013	0.10013	9.9112	789.74	7887.4	78174.	70
80	0.00049	9.9951	99.561	0.00005	0.10005	9.9609	2048.0	20474.	203940.	80
90	0.00019	9.9981	99.811	0.00002	0.10002	9.9830	5313.0	53120.	530302.	90
100	0.00007	9.9992	99.920	0.00001	0.10001	9.9927	13780.	137796.	>10**6	100
120	0.00001	9.9998	99.986	0.00000	0.10000	9.9987	92709.	927081.	>10**6	120
INF	0.00000	10.000	130.00	0.0000	0.10000	10.000	INF	INF	INF	INF

| | $i=11\%$ | | | $i=11\%$ | | | $i=11\%$ | | | |
| | PRESENT SUM, P | | | UNIFORM SERIES, A | | | FUTURE SUM, F | | | |
n	P/F	P/A	P/G	A/F	A/P	A/G	F/P	F/A	F/G	n
1	0.90090	0.9009	0.0000	1.00000	1.11000	0.0000	1.1100	1.0000	0.0000	1
2	0.81162	1.7125	0.8116	0.47393	0.53393	0.4739	1.2321	2.1100	1.0000	2
3	0.73119	2.4437	2.2740	0.29921	0.40921	0.9305	1.3676	3.3421	3.1100	3
4	0.65873	3.1024	4.2502	0.21233	0.32233	1.3699	1.5180	4.7097	6.4521	4
5	0.59345	3.6959	6.6240	0.16057	0.27057	1.7922	1.6850	6.2278	11.161	5
6	0.53464	4.2305	9.2972	0.12638	0.23638	2.1976	1.8704	7.9128	17.389	6
7	0.48166	4.7122	12.187	0.10222	0.21222	2.5863	2.0761	9.7832	25.302	7
8	0.43393	5.1461	15.224	0.08432	0.19432	2.9584	2.3045	11.859	35.085	8
9	0.39092	5.5370	18.352	0.07060	0.18060	3.3144	2.5580	14.164	46.945	9
10	0.35218	5.8892	21.521	0.05980	0.16980	3.6544	2.8394	16.722	61.109	10
11	0.31728	6.2065	24.694	0.05112	0.16112	3.9788	3.1517	19.561	77.831	11
12	0.28584	6.4923	27.838	0.04403	0.15403	4.2879	3.4984	22.713	97.392	12
13	0.25751	6.7498	30.929	0.03815	0.14815	4.5821	3.8832	26.211	120.10	13
14	0.23199	6.9818	33.944	0.03323	0.14323	4.8618	4.3104	30.094	146.31	14
15	0.20900	7.1908	35.870	0.02907	0.13907	5.1274	4.7845	34.405	176.41	15
16	0.18829	7.3791	39.695	0.02552	0.13552	5.3793	5.3108	39.189	210.81	16
17	0.16963	7.5487	42.409	0.02247	0.13247	5.6180	5.8950	44.500	250.00	17
18	0.15282	7.7016	45.007	0.01984	0.12984	5.8438	6.5435	50.395	294.50	18
19	0.13768	7.8392	47.485	0.01756	0.12756	6.0573	7.2633	56.939	344.90	19
20	0.12403	7.9633	49.842	0.01558	0.12558	6.2589	8.0623	64.202	401.84	20
21	0.11174	8.0750	52.077	0.01384	0.12384	6.4491	8.9491	72.265	466.04	21
22	0.10067	8.1757	54.191	0.01231	0.12231	6.6282	9.9335	81.214	538.31	22
23	0.09069	8.2664	56.186	0.01097	0.12097	6.7969	11.026	91.147	619.52	23
24	0.08170	8.3481	58.065	0.00979	0.11979	6.9555	12.239	102.17	710.67	24
25	0.07361	8.4217	59.832	0.00874	0.11874	7.1044	13.585	114.41	812.84	25
26	0.06631	8.4880	61.490	0.00781	0.11781	7.2443	15.079	127.99	927.26	26
27	0.05974	8.5478	63.043	0.00699	0.11699	7.3753	16.738	143.07	1055.2	27
28	0.05382	8.6016	64.496	0.00626	0.11626	7.4981	18.579	159.81	1198.3	28
29	0.04849	8.6501	65.854	0.00561	0.11561	7.6131	20.623	178.39	1358.1	29
30	0.04368	8.6937	67.121	0.00502	0.11502	7.7205	22.892	199.02	1536.5	30
35	0.02592	8.8552	72.253	0.00293	0.11293	8.1594	38.574	341.59	2787.1	35
40	0.01538	8.9510	75.778	0.00172	0.11172	8.4659	65.000	581.82	4925.6	40
45	0.00913	9.0079	78.155	0.00101	0.11101	8.6762	109.53	986.63	8550.3	45
50	0.00542	9.0416	79.734	0.00060	0.11060	8.8185	184.56	1668.7	14716.	50
55	0.00322	9.0616	80.771	0.00035	0.11035	8.9134	311.00	2818.2	25120.	55
60	0.00191	9.0735	81.446	0.00021	0.11021	8.9762	524.05	4755.0	42682.	60
65	0.00113	9.0806	81.881	0.00013	0.11013	9.0172	883.06	8018.7	72307.	65
70	0.00067	9.0848	82.161	0.00007	0.11007	9.0438	1488.0	13518.	122258.	70
INF	0.00000	9.0909	82.644	0.0000	0.11000	9.0909	INF	INF	INF	INF

	$i = 12\%$ PRESENT SUM, P			$i = 12\%$ UNIFORM SERIES, A			$i = 12\%$ FUTURE SUM, F			
n	P/F	P/A	P/G	A/F	A/P	A/G	F/P	F/A	F/G	n
1	0.89286	0.8928	0.0000	1.00000	1.12000	0.0000	1.1200	1.0000	0.0000	1
2	0.79719	1.6900	0.7971	0.47170	0.59170	0.4717	1.2544	2.1200	1.0000	2
3	0.71178	2.4018	2.2207	0.29635	0.41635	0.9246	1.4049	3.3744	3.1200	3
4	0.63552	3.0373	4.1273	0.20923	0.32923	1.3588	1.5735	4.7793	6.4944	4
5	0.56743	3.6047	6.3970	0.15741	0.27741	1.7745	1.7623	6.3528	11.273	5
6	0.50663	4.1114	8.9301	0.12323	0.24323	2.1720	1.9738	8.1151	17.626	6
7	0.45235	4.5637	11.644	0.09912	0.21912	2.5514	2.2106	10.089	25.741	7
8	0.40388	4.9676	14.471	0.08130	0.20130	2.9131	2.4759	12.299	35.830	8
9	0.36061	5.3282	17.356	0.06768	0.18768	3.2574	2.7730	14.775	48.130	9
10	0.32197	5.6502	20.254	0.05698	0.17698	3.5846	3.1058	17.548	62.906	10
11	0.28748	5.9377	23.128	0.04842	0.16842	3.8952	3.4785	20.654	80.454	11
12	0.25668	6.1943	25.952	0.04144	0.16144	4.1896	3.8959	24.133	101.10	12
13	0.22917	6.4235	28.702	0.03568	0.15568	4.4683	4.3634	28.029	125.24	13
14	0.20462	6.6281	31.362	0.03087	0.15087	4.7316	4.8871	32.392	153.27	14
15	0.18270	6.8108	33.920	0.02682	0.14682	4.9803	5.4735	37.279	185.66	15
16	0.16312	6.9739	36.367	0.02339	0.14339	5.2146	6.1303	42.753	222.94	16
17	0.14564	7.1196	38.697	0.02046	0.14046	5.4353	6.8660	48.883	265.69	17
18	0.13004	7.2496	40.909	0.01794	0.13794	5.6427	7.6899	55.749	314.58	18
19	0.11611	7.3657	42.997	0.01576	0.13576	5.8375	8.6127	63.439	370.33	19
20	0.10367	7.4694	44.967	0.01388	0.13388	6.0202	9.6462	72.052	433.77	20
21	0.09256	7.5620	46.818	0.01224	0.13224	6.1913	10.803	81.698	505.82	21
22	0.08264	7.6446	48.554	0.01081	0.13081	6.3514	12.100	92.502	587.52	22
23	0.07379	7.7184	50.177	0.00956	0.12956	6.5010	13.552	104.60	680.02	23
24	0.06588	7.7843	51.692	0.00846	0.12846	6.6406	15.178	118.15	784.62	24
25	0.05882	7.8431	53.104	0.00750	0.12750	6.7708	17.000	133.33	902.78	25
26	0.05252	7.8956	54.417	0.00665	0.12665	6.8921	19.040	150.33	1036.1	26
27	0.04689	7.9425	55.636	0.00590	0.12590	7.0097	21.324	169.37	1186.4	27
28	0.04187	7.9844	56.767	0.00524	0.12524	7.1097	23.883	190.69	1355.3	28
29	0.03738	8.0218	57.814	0.00466	0.12466	7.2071	26.749	214.58	1546.5	29
30	0.03338	8.0551	58.782	0.00414	0.12414	7.2974	29.959	241.33	1761.1	30
35	0.01894	8.1755	62.605	0.00232	0.12232	7.6576	52.799	431.66	3305.5	35
40	0.01075	8.2437	65.115	0.00130	0.12130	7.8987	93.051	767.09	6059.1	40
45	0.00610	8.2825	66.734	0.00074	0.12074	8.0572	163.98	1358.2	10943.	45
50	0.00346	8.3045	67.762	0.00042	0.12042	8.1597	289.00	2400.0	19583.	50
55	0.00196	8.3169	68.408	0.00024	0.12024	8.2251	509.32	4236.0	34841.	55
60	0.00111	8.3240	68.810	0.00013	0.12013	8.2664	897.59	7471.6	61763.	60
65	0.00063	8.3280	69.058	0.00008	0.12008	8.2922	1581.8	13173.	109241.	65
70	0.00036	8.3303	69.210	0.00004	0.12004	8.3082	2787.8	23223.	192944.	70
INF	0.00000	8.3333	69.444	0.0000	0.12000	8.3333	INF	INF	INF	INF

		i = 13%			i = 13%			i = 13%		
		PRESENT SUM, P			UNIFORM SERIES, A			FUTURE SUM, F		
n	P/F	P/A	P/G	A/F	A/P	A/G	F/P	F/A	F/G	n
1	0.88496	0.8849	0.0000	1.00000	1.13000	0.0000	1.1300	1.0000	0.0000	1
2	0.78315	1.6681	0.7831	0.46948	0.59948	0.4694	1.2769	2.1300	1.0000	2
3	0.69305	2.3611	2.1692	0.29352	0.42352	0.9187	1.4429	3.4069	3.1300	3
4	0.61332	2.9744	4.0092	0.20619	0.33619	1.3478	1.6304	4.8498	6.5369	4
5	0.54276	3.5172	6.1802	0.15431	0.28431	1.7571	1.8424	6.4802	11.386	5
6	0.48032	3.9975	8.5818	0.12015	0.25015	2.1467	2.0819	8.3227	17.867	6
7	0.42506	4.4226	11.132	0.09611	0.22611	2.5171	2.3526	10.404	26.189	7
8	0.37616	4.7987	13.765	0.07839	0.20839	2.8685	2.6584	12.757	36.594	8
9	0.33288	5.1316	16.428	0.06487	0.19487	3.2013	3.0040	15.415	49.351	9
10	0.29459	5.4262	19.079	0.05429	0.18429	3.5161	3.3945	18.419	64.767	10
11	0.26070	5.6869	21.686	0.04584	0.17584	3.8134	3.8358	21.814	83.187	11
12	0.23071	5.9176	24.224	0.03899	0.16899	4.0935	4.3345	25.650	105.00	12
13	0.20416	6.1218	26.674	0.03335	0.16335	4.3572	4.8980	29.984	130.65	13
14	0.18068	6.3024	29.023	0.02867	0.15867	4.6050	5.5347	34.882	160.63	14
15	0.15989	6.4623	31.261	0.02474	0.15474	4.8374	6.2542	40.417	195.51	15
16	0.14150	6.6038	33.384	0.02143	0.15143	5.0552	7.0673	46.671	235.93	16
17	0.12522	6.7290	35.387	0.01861	0.14861	5.2589	7.9860	53.739	282.60	17
18	0.11081	6.8399	37.271	0.01520	0.14620	5.4491	9.0242	61.725	336.34	18
19	0.09806	6.9379	39.036	0.01413	0.14413	5.6265	10.197	70.749	398.07	19
20	0.08678	7.0247	40.685	0.01235	0.14235	5.7917	11.523	80.946	468.82	20
21	0.07680	7.1015	42.221	0.01081	0.14081	5.9453	13.021	92.469	549.76	21
22	0.06796	7.1695	43.648	0.00948	0.13948	6.0880	14.713	105.49	642.23	22
23	0.06014	7.2296	44.971	0.00832	0.13832	6.2204	16.626	120.20	747.73	23
24	0.05323	7.2828	46.196	0.00731	0.13731	6.3430	18.788	136.83	867.93	24
25	0.04710	7.3299	47.326	0.00643	0.13643	6.4565	21.230	155.62	1004.7	25
26	0.04168	7.3716	48.368	0.00565	0.13565	6.5614	23.990	176.85	1160.3	26
27	0.03689	7.4085	49.327	0.00498	0.13498	6.6581	27.109	200.84	1337.2	27
28	0.03264	7.4412	50.209	0.00439	0.13439	6.7474	30.633	227.95	1538.0	28
29	0.02889	7.4700	51.017	0.00387	0.13387	6.8296	34.615	258.58	1766.0	29
30	0.02557	7.4956	51.759	0.00341	0.13341	6.9052	39.115	293.19	2024.6	30
35	0.01388	7.5855	54.614	0.00183	0.13183	7.1998	72.068	546.68	3936.0	35
40	0.00753	7.6343	56.408	0.00099	0.13099	7.3887	132.78	1013.7	7490.0	40
45	0.00409	7.6608	57.514	0.00053	0.13053	7.5076	244.64	1874.1	14070.	45
50	0.00222	7.6752	58.187	0.00029	0.13029	7.5811	450.73	3459.5	26227.	50
55	0.00120	7.6830	58.590	0.00016	0.13016	7.6260	830.45	6380.4	48656.	55
60	0.00065	7.6872	58.831	0.00009	0.13009	7.6530	1530.0	11761.	90015.	60
65	0.00035	7.6895	58.973	0.00005	0.13005	7.6692	2819.0	21677.	166247.	65
70	0.00019	7.6908	59.056	0.00003	0.13003	7.6788	5193.8	39945.	306732.	70
INF	0.00000	7.6923	59.171	0.0000	0.13000	7.6923	INF	INF	INF	INF

i = 14%

n	P/F	P/A	P/G	A/F	A/P	A/G	F/P	F/A	F/G	n
1	0.87719	0.8771	0.0000	1.00000	1.14000	0.0000	1.1400	1.0000	0.0000	1
2	0.76947	1.6466	0.7694	0.46729	0.60729	0.4672	1.2996	2.1400	1.0000	2
3	0.67497	2.3216	2.1194	0.29073	0.43073	0.9129	1.4815	3.4396	3.1400	3
4	0.59208	2.9137	3.8956	0.20320	0.34320	1.3370	1.6889	4.9211	6.5796	4
5	0.51937	3.4330	5.9731	0.15128	0.29128	1.7398	1.9254	6.6101	11.500	5
6	0.45559	3.8886	8.2510	0.11716	0.25716	2.1218	2.1949	8.5355	18.110	6
7	0.39964	4.2883	10.648	0.09319	0.23319	2.4832	2.5022	10.730	26.646	7
8	0.35056	4.6388	13.102	0.07557	0.21557	2.8245	2.8525	13.232	37.376	8
9	0.30751	4.9463	15.562	0.06217	0.20217	3.1463	3.2519	16.085	50.609	9
10	0.26974	5.2161	17.990	0.05171	0.19171	3.4490	3.7072	19.337	66.695	10
11	0.23662	5.4527	20.356	0.04339	0.18339	3.7333	4.2262	23.044	86.032	11
12	0.20756	5.6602	22.639	0.03667	0.17667	3.9997	4.8179	27.270	109.07	12
13	0.18207	5.8423	24.824	0.03116	0.17116	4.2490	5.4924	32.088	136.34	13
14	0.15971	6.0023	26.900	0.02661	0.16661	4.4819	6.2613	37.581	168.43	14
15	0.14010	6.1421	28.862	0.02281	0.16281	4.6990	7.1379	43.842	206.01	15
16	0.12289	6.2650	30.705	0.01962	0.15962	4.9011	8.1372	50.980	249.86	16
17	0.10780	6.3728	32.430	0.01692	0.15692	5.0888	9.2764	59.117	300.84	17
18	0.09456	6.4674	34.038	0.01462	0.15462	5.2629	10.575	68.394	359.95	18
19	0.08295	6.5503	35.531	0.01266	0.15266	5.4242	12.055	78.969	428.35	19
20	0.07276	6.6231	36.913	0.01099	0.15099	5.5734	13.743	91.024	507.32	20
21	0.06383	6.6869	38.190	0.00954	0.14954	5.7111	15.667	104.76	598.34	21
22	0.05599	6.7429	39.365	0.00830	0.14830	5.8380	17.861	120.43	703.11	22
23	0.04911	6.7920	40.446	0.00723	0.14723	5.9549	20.361	138.29	823.55	23
24	0.04308	6.8351	41.437	0.00630	0.14630	6.0623	23.212	158.65	961.84	24
25	0.03779	6.8729	42.344	0.00550	0.14550	6.1610	26.461	181.87	1120.5	25
26	0.03315	6.9060	43.172	0.00480	0.14480	6.2514	30.166	208.33	1302.3	26
27	0.02908	6.9351	43.928	0.00419	0.14419	6.3342	34.389	238.49	1510.7	27
28	0.02551	6.9606	44.617	0.00366	0.14366	6.4099	39.204	272.88	1749.2	28
29	0.02237	6.9830	45.244	0.00320	0.14320	6.4791	44.693	312.09	2022.1	29
30	0.01963	7.0026	45.813	0.00280	0.14280	6.5422	50.950	356.78	2334.1	30
35	0.01019	7.0700	47.951	0.00144	0.14144	6.7824	98.100	693.57	4704.0	35
40	0.00529	7.1050	49.237	0.00075	0.14075	6.9299	188.88	1342.0	9300.1	40
45	0.00275	7.1232	49.996	0.00039	0.14039	7.0187	363.67	2590.5	18182.	45
50	0.00143	7.1326	50.437	0.00020	0.14020	7.0713	700.23	4994.5	35318.	50
55	0.00074	7.1375	50.691	0.00010	0.14010	7.1020	1348.2	9623.1	68343.	55
60	0.00039	7.1401	50.835	0.00005	0.14005	7.1197	2595.9	18535.	131965.	60
65	0.00020	7.1414	50.917	0.00002	0.14003	7.1298	4998.2	35694.	254496.	65
70	0.00010	7.1421	50.963	0.00001	0.14001	7.1355	9623.6	68733.	490451.	70
INF	0.00000	7.1428	51.020	0.0000	0.14000	7.1428	INF	INF	INF	INF

PRESENT SUM, P — UNIFORM SERIES, A — FUTURE SUM, F

PRESENT SUM, P

n	P/F	P/A	P/G
1	0.86957	0.8695	0.0000
2	0.75614	1.6257	0.7561
3	0.65752	2.2832	2.0711
4	0.57175	2.8549	3.7864
5	0.49718	3.3521	5.7751
6	0.43233	3.7844	7.9367
7	0.37594	4.1604	10.192
8	0.32690	4.4873	12.480
9	0.28426	4.7715	14.754
10	0.24718	5.0187	16.979
11	0.21494	5.2337	19.128
12	0.18691	5.4206	21.184
13	0.16253	5.5831	23.135
14	0.14133	5.7244	24.972
15	0.12289	5.8473	26.693
16	0.10686	5.9542	28.296
17	0.09293	6.0471	29.782
18	0.08081	6.1279	31.156
19	0.07027	6.1982	32.421
20	0.06110	6.2593	33.582
21	0.05313	6.3124	34.644
22	0.04620	6.3586	35.615
23	0.04017	6.3988	36.498
24	0.03493	6.4337	37.302
25	0.03038	6.4641	38.031
26	0.02642	6.4905	38.691
27	0.02297	6.5135	39.289
28	0.01997	6.5335	39.828
29	0.01737	6.5508	40.314
30	0.01510	6.5659	40.752
35	0.00751	6.6166	42.358
40	0.00373	6.6417	43.283
45	0.00186	6.6542	43.805
50	0.00092	6.6605	44.095
55	0.00046	6.6636	44.255
60	0.00023	6.6651	44.343
65	0.00011	6.6659	44.390
INF	0.00000	6.6666	44.444

UNIFORM SERIES, A ; FUTURE SUM, F

n	A/F	A/P	A/G	F/P	F/A	F/G
1	1.00000	1.15000	0.0000	1.1500	1.0000	0.0000
2	0.46512	0.61512	0.4651	1.3225	2.1500	1.0000
3	0.28798	0.43798	0.9071	1.5208	3.4725	3.1500
4	0.20027	0.35027	1.3262	1.7490	4.9933	6.6225
5	0.14832	0.29832	1.7228	2.0113	6.7423	11.615
6	0.11424	0.26424	2.0971	2.3130	8.7537	18.358
7	0.09036	0.24036	2.4498	2.6600	11.066	27.112
8	0.07285	0.22285	2.7813	3.0590	13.726	38.178
9	0.05957	0.20957	3.0922	3.5178	16.785	51.905
10	0.04925	0.19925	3.3832	4.0455	20.303	68.691
11	0.04107	0.19107	3.6549	4.6523	24.349	88.995
12	0.03448	0.18448	3.9082	5.3502	29.001	113.34
13	0.02911	0.17911	4.1437	6.1527	34.351	142.34
14	0.02469	0.17469	4.3624	7.0757	40.504	176.69
15	0.02102	0.17102	4.5649	8.1370	47.580	217.20
16	0.01795	0.16795	4.7522	9.3576	55.717	264.78
17	0.01537	0.16537	4.9250	10.761	65.075	320.50
18	0.01319	0.16319	5.0843	12.375	75.836	385.57
19	0.01134	0.16134	5.2307	14.231	88.211	461.41
20	0.00976	0.15976	5.3651	16.366	102.44	549.62
21	0.00842	0.15842	5.4883	18.821	118.81	652.06
22	0.00727	0.15727	5.6010	21.644	137.63	770.87
23	0.00628	0.15628	5.7039	24.891	159.27	908.50
24	0.00543	0.15543	5.7978	28.625	184.16	1067.7
25	0.00470	0.15470	5.8834	32.919	212.79	1251.9
26	0.00407	0.15407	5.9612	37.856	245.71	1464.7
27	0.00353	0.15353	6.0319	43.535	283.56	1710.4
28	0.00306	0.15306	6.0960	50.065	327.10	1994.0
29	0.00265	0.15265	6.1540	57.575	377.17	2321.1
30	0.00230	0.15230	6.2066	66.211	434.74	2698.3
35	0.00113	0.15113	6.4018	133.17	881.17	5641.1
40	0.00056	0.15056	6.5167	267.86	1779.0	11593.
45	0.00028	0.15028	6.5829	538.76	3585.1	23600.
50	0.00014	0.15014	6.6204	1083.6	7217.7	47784.
55	0.00007	0.15007	6.6414	2179.6	14524.	96461.
60	0.00003	0.15003	6.6529	4384.0	29220.	194400.
65	0.00002	0.15002	6.6592	8817.7	58778.	391424.
INF	0.0000	0.15000	6.6666	INF	INF	INF

n	P/F	P/A	P/G	A/F	A/P	A/G	F/P	F/A	F/G	n
	PRESENT SUM, P			UNIFORM SERIES, A			FUTURE SUM, F			
1	0.83333	0.8333	0.0000	1.00000	1.20000	0.0000	1.2000	1.0000	0.0000	1
2	0.69444	1.5277	0.6944	0.45455	0.65455	0.4545	1.4400	2.2000	1.0000	2
3	0.57870	2.1064	1.8518	0.27473	0.47473	0.8791	1.7280	3.6400	3.2000	3
4	0.48225	2.5887	3.2986	0.18629	0.38629	1.2742	2.0736	5.3680	6.8400	4
5	0.40188	2.9906	4.9061	0.13438	0.33438	1.6405	2.4883	7.4416	12.208	5
6	0.33490	3.3255	6.5806	0.10071	0.30071	1.9788	2.9859	9.9299	19.649	6
7	0.27908	3.6045	8.2551	0.07742	0.27742	2.2901	3.5831	12.915	29.579	7
8	0.23257	3.8371	9.8830	0.06061	0.26061	2.5756	4.2998	16.499	42.495	8
9	0.19381	4.0309	11.433	0.04808	0.24808	2.8364	5.1597	20.798	58.994	9
10	0.16151	4.1924	12.887	0.03852	0.23852	3.0738	6.1917	25.958	79.793	10
11	0.13459	4.3270	14.233	0.03110	0.23110	3.2892	7.4300	32.150	105.75	11
12	0.11216	4.4392	15.466	0.02526	0.22526	3.4841	8.9161	39.580	137.90	12
13	0.09346	4.5326	16.588	0.02062	0.22062	3.6597	10.699	48.496	177.48	13
14	0.07789	4.6105	17.600	0.01689	0.21689	3.8174	12.839	59.195	225.98	14
15	0.06491	4.6754	18.509	0.01388	0.21388	3.9588	15.407	72.035	285.17	15
16	0.05409	4.7295	19.320	0.01144	0.21144	4.0851	18.488	87.442	357.21	16
17	0.04507	4.7746	20.041	0.00944	0.20944	4.1975	22.186	105.93	444.65	17
18	0.03756	4.8121	20.680	0.00781	0.20781	4.2975	26.623	128.11	550.58	18
19	0.03130	4.8435	21.243	0.00646	0.20646	4.3860	31.948	154.74	678.70	19
20	0.02608	4.8695	21.739	0.00536	0.20536	4.4643	38.337	186.68	833.44	20
21	0.02174	4.8913	22.174	0.00444	0.20444	4.5333	46.005	225.02	1020.1	21
22	0.01811	4.9094	22.554	0.00369	0.20369	4.5941	55.206	271.03	1245.1	22
23	0.01509	4.9245	22.886	0.00307	0.20307	4.6475	66.247	326.23	1516.1	23
24	0.01258	4.9371	23.176	0.00255	0.20255	4.6942	79.496	392.48	1842.4	24
25	0.01048	4.9475	23.427	0.00212	0.20212	4.7351	95.396	471.98	2234.9	25
26	0.00874	4.9563	23.646	0.00176	0.20176	4.7708	114.47	567.37	2706.8	26
27	0.00728	4.9636	23.835	0.00147	0.20147	4.8020	137.37	681.85	3274.2	27
28	0.00607	4.9696	23.999	0.00122	0.20122	4.8291	164.84	819.22	3956.1	28
29	0.00506	4.9747	24.140	0.00102	0.20102	4.8526	197.81	984.06	4775.3	29
30	0.00421	4.9789	24.262	0.00085	0.20085	4.8730	237.37	1181.8	5759.4	30
35	0.00169	4.9915	24.661	0.00034	0.20034	4.9406	590.66	2948.3	14566.	35
40	0.00068	4.9966	24.846	0.00014	0.20014	4.9876	1469.7	7343.8	36519.	40
45	0.00027	4.9986	24.931	0.00005	0.20005	4.9945	3657.2	18281.	91181.	45
50	0.00011	4.9994	24.969	0.00002	0.20002	4.9975	9100.4	45497.	227236.	50
55	0.00004	4.9997	24.986	0.00001	0.20001	4.9995	22644.	113219.	565820.	55
60	0.00002	4.9999	24.994	0.00000	0.20000	4.9989	56347.	281733	>10**6	60
INF	0.00000	5.0000	25.000	0.0000	0.20000	5.0000	INF	INF	INF	INF

i = 25%

n	P/F	P/A	P/G	A/F	A/P	A/G	F/P	F/A	F/G	n
	PRESENT SUM, P			UNIFORM SERIES, A			FUTURE SUM, F			
1	0.80000	0.8000	0.0000	1.00000	1.25000	0.0000	1.2500	1.0000	0.0000	1
2	0.64000	1.4400	0.6400	0.44444	0.69444	0.4444	1.5625	2.2500	1.0000	2
3	0.51200	1.9520	1.6640	0.26230	0.51230	0.8524	1.9531	3.8125	3.2500	3
4	0.40960	2.3616	2.8928	0.17344	0.42344	1.2249	2.4414	5.7656	7.0625	4
5	0.32768	2.6892	4.2035	0.12185	0.37185	1.5630	3.0517	8.2070	12.828	5
6	0.26214	2.9514	5.5142	0.08882	0.33882	1.8683	3.8147	11.258	21.035	6
7	0.20972	3.1611	6.7725	0.06634	0.31634	2.1424	4.7683	15.073	32.293	7
8	0.16777	3.3289	7.9469	0.05040	0.30040	2.3872	5.9604	19.841	47.367	8
9	0.13422	3.4631	9.0206	0.03876	0.28876	2.6047	7.4505	25.802	67.209	9
10	0.10737	3.5705	9.9870	0.03007	0.28007	2.7971	9.3132	33.252	93.011	10
11	0.08590	3.6564	10.846	0.02349	0.27349	2.9663	11.641	42.566	126.26	11
12	0.06872	3.7251	11.602	0.01845	0.26845	3.1145	14.551	54.207	168.83	12
13	0.05498	3.7801	12.261	0.01454	0.26454	3.2437	18.189	68.759	223.03	13
14	0.04398	3.8240	12.833	0.01150	0.26150	3.3559	22.737	86.949	291.79	14
15	0.03518	3.8592	13.326	0.00912	0.25912	3.4529	28.421	109.68	378.74	15
16	0.02815	3.8874	13.748	0.00724	0.25724	3.5366	35.527	138.10	488.43	16
17	0.02252	3.9099	14.108	0.00576	0.25576	3.6083	44.408	173.63	626.54	17
18	0.01801	3.9279	14.414	0.00459	0.25459	3.6697	55.511	218.04	800.17	18
19	0.01441	3.9423	14.674	0.00366	0.25366	3.7221	69.388	273.55	1018.2	19
20	0.01153	3.9538	14.893	0.00292	0.25292	3.7667	86.736	342.94	1291.7	20
21	0.00922	3.9631	15.077	0.00233	0.25233	3.8045	108.42	429.68	1634.7	21
22	0.00738	3.9704	15.232	0.00186	0.25186	3.8364	135.52	538.10	2064.4	22
23	0.00590	3.9763	15.362	0.00148	0.25148	3.8634	169.40	673.62	2602.5	23
24	0.00472	3.9811	15.471	0.00119	0.25119	3.8861	211.75	843.03	3276.1	24
25	0.00378	3.9848	15.561	0.00095	0.25095	3.9051	264.69	1054.7	4119.1	25
26	0.00302	3.9879	15.637	0.00076	0.25076	3.9211	330.87	1319.4	5173.9	26
27	0.00242	3.9903	15.700	0.00061	0.25061	3.9345	413.59	1650.3	6493.4	27
28	0.00193	3.9922	15.752	0.00048	0.25048	3.9457	516.98	2063.9	8143.8	28
29	0.00155	3.9938	15.795	0.00039	0.25039	3.9550	646.23	2580.9	10207.	29
30	0.00124	3.9950	15.831	0.00031	0.25031	3.9628	807.79	3227.1	12788.	30
35	0.00041	3.9983	15.936	0.00010	0.25010	3.9858	2465.1	9856.7	39287.	35
40	0.00013	3.9994	15.976	0.00003	0.25003	3.9946	7523.1	30088.	120195.	40
45	0.00004	3.9998	15.991	0.00001	0.25001	3.9980	22958.	91831.	367146.	45
INF	0.00000	4.0000	16.000	0.0000	0.25000	4.0000	INF	INF	INF	INF

| | i = 30% | | | i = 30% | | | i = 30% | | | |
| | PRESENT SUM, P | | | UNIFORM SERIES, A | | | FUTURE SUM, F | | | |
n	P/F	P/A	P/G	A/F	A/P	A/G	F/P	F/A	F/G	n
1	0.76923	0.7692	0.0000	1.00000	1.30000	0.0000	1.3000	1.0000	0.0000	1
2	0.59172	1.3609	0.5917	0.43478	0.73478	0.4347	1.6900	2.3000	1.0000	2
3	0.45517	1.8161	1.5020	0.25063	0.55063	0.8270	2.1970	3.9900	3.3000	3
4	0.35013	2.1662	2.5524	0.16163	0.46163	1.1782	2.8561	6.1870	7.2900	4
5	0.26933	2.4355	3.6297	0.11058	0.41058	1.4903	3.7129	9.0431	13.477	5
6	0.20718	2.6427	4.6656	0.07839	0.37839	1.7654	4.8268	12.756	22.520	6
7	0.15937	2.8021	5.6218	0.05687	0.35687	2.0062	6.2748	17.582	35.276	7
8	0.12259	2.9247	6.4799	0.04192	0.34192	2.2155	8.1573	23.857	52.859	8
9	0.09430	3.0190	7.2343	0.03124	0.33124	2.3962	10.604	32.015	76.716	9
10	0.07254	3.0915	7.8871	0.02346	0.32346	2.5512	13.785	42.619	108.73	10
11	0.05580	3.1473	8.4451	0.01773	0.31773	2.6832	17.921	56.405	151.35	11
12	0.04292	3.1902	8.9173	0.01345	0.31345	2.7951	23.298	74.327	207.75	12
13	0.03302	3.2232	9.3135	0.01024	0.31024	2.8894	30.287	97.625	282.08	13
14	0.02540	3.2486	9.6436	0.00782	0.30782	2.9685	39.373	127.91	379.70	14
15	0.01954	3.2682	9.9172	0.00598	0.30598	3.0344	51.185	167.28	507.62	15
16	0.01503	3.2832	10.142	0.00458	0.30458	3.0892	66.541	218.47	674.90	16
17	0.01156	3.2948	10.327	0.00351	0.30351	3.1345	86.504	285.01	893.38	17
18	0.00889	3.3036	10.478	0.00269	0.30269	3.1718	112.45	371.51	1178.3	18
19	0.00684	3.3105	10.601	0.00207	0.30207	3.2024	146.19	483.97	1549.9	19
20	0.00526	3.3157	10.701	0.00159	0.30159	3.2275	190.05	630.16	2033.8	20
21	0.00405	3.3198	10.782	0.00122	0.30122	3.2479	247.06	820.21	2664.0	21
22	0.00311	3.3229	10.848	0.00094	0.30094	3.2646	321.18	1067.2	3484.2	22
23	0.00239	3.3253	10.900	0.00072	0.30072	3.2781	417.53	1388.4	4551.5	23
24	0.00184	3.3271	10.943	0.00055	0.30055	3.2890	542.80	1806.0	5940.0	24
25	0.00142	3.3286	10.977	0.00043	0.30043	3.2978	705.64	2348.8	7746.0	25
26	0.00109	3.3297	11.004	0.00033	0.30033	3.3049	917.33	3054.4	10094.	26
27	0.00084	3.3305	11.026	0.00025	0.30025	3.3106	1192.5	3971.7	13149.	27
28	0.00065	3.3311	11.043	0.00019	0.30019	3.3152	1550.2	5164.3	17121.	28
29	0.00050	3.3316	11.057	0.00015	0.30015	3.3189	2015.3	6714.6	22285.	29
30	0.00038	3.3320	11.068	0.00011	0.30011	3.3218	2620.0	8729.9	29000.	30
35	0.00010	3.3329	11.098	0.00003	0.30003	3.3297	9727.8	32422.	107960.	35
INF	0.00000	3.3333	11.111	0.0000	0.30000	3.3333	INF	INF	INF	INF

289

i = 40%

PRESENT SUM, P

n	P/F	P/A	P/G
1	0.71429	0.7142	0.0000
2	0.51020	1.2244	0.5102
3	0.36443	1.5889	1.2390
4	0.26031	1.8492	2.0199
5	0.18593	2.0351	2.7637
6	0.13281	2.1679	3.4277
7	0.09486	2.2628	3.9969
8	0.06776	2.3306	4.4712
9	0.04840	2.3790	4.8584
10	0.03457	2.4135	5.1696
11	0.02469	2.4382	5.4165
12	0.01764	2.4559	5.6106
13	0.01260	2.4685	5.7617
14	0.00900	2.4775	5.8787
15	0.00643	2.4839	5.9687
16	0.00459	2.4885	6.0376
17	0.00328	2.4918	6.0901
18	0.00234	2.4941	6.1299
19	0.00167	2.4958	6.1600
20	0.00120	2.4970	6.1827
21	0.00085	2.4978	6.1998
22	0.00061	2.4984	6.2126
23	0.00044	2.4989	6.2222
24	0.00031	2.4992	6.2293
25	0.00022	2.4994	6.2347
INF	0.00000	2.5000	6.2500

i = 40%

UNIFORM SERIES, A

A/F	A/P	A/G
1.00000	1.40000	0.0000
0.41667	0.81667	0.4166
0.22936	0.62936	0.7798
0.14077	0.54077	1.0923
0.09136	0.49136	1.3579
0.06126	0.46126	1.5811
0.04192	0.44192	1.7663
0.02907	0.42907	1.9185
0.02034	0.42034	2.0422
0.01432	0.41432	2.1419
0.01013	0.41013	2.2214
0.00718	0.40718	2.2845
0.00510	0.40510	2.3341
0.00363	0.40363	2.3728
0.00259	0.40259	2.4029
0.00185	0.40185	2.4262
0.00132	0.40132	2.4440
0.00094	0.40094	2.4577
0.00067	0.40067	2.4681
0.00048	0.40048	2.4760
0.00034	0.40034	2.4820
0.00024	0.40024	2.4865
0.00017	0.40017	2.4899
0.00012	0.40012	2.4925
0.00009	0.40009	2.4944
0.0000	0.40000	2.5000

i = 40%

FUTURE SUM, F

F/P	F/A	F/G	n
1.4000	1.0000	0.0000	1
1.9600	2.4000	1.0000	2
2.7440	4.3600	3.4000	3
3.8416	7.1040	7.7600	4
5.3782	10.945	14.864	5
7.5295	16.323	25.809	6
10.541	23.853	42.133	7
14.757	34.394	65.986	8
20.661	49.152	100.38	9
28.925	69.813	149.53	10
40.495	98.739	219.34	11
56.693	139.23	318.08	12
79.371	195.92	457.32	13
111.12	275.30	653.25	14
155.56	386.42	928.55	15
217.79	541.98	1314.9	16
304.91	759.78	1856.9	17
426.87	1064.7	2616.7	18
597.63	1491.5	3681.4	19
836.68	2089.2	5173.0	20
1171.3	2925.8	7262.2	21
1639.9	4097.2	10188.	22
2295.8	5737.1	14285.	23
3214.2	8033.0	20022.	24
4499.8	11247.	28055.	25
INF	INF	INF	INF

	$i=50\%$	PRESENT SUM, P			$i=50\%$	UNIFORM SERIES, A			$i=50\%$	FUTURE SUM, F		
n	P/F	P/A	P/G	A/F	A/P	A/G	F/P	F/A	F/G	n
1	0.66667	0.6666	0.0000	1.00000	1.50000	0.0000	1.5000	1.0000	0.0000	1
2	0.44444	1.1111	0.4444	0.40000	0.90000	0.4000	2.2500	2.5000	1.0000	2
3	0.29630	1.4074	1.0370	0.21053	0.71053	0.7368	3.3750	4.7500	3.5000	3
4	0.19753	1.6049	1.6296	0.12308	0.62308	1.0153	5.0625	8.1250	8.2500	4
5	0.13169	1.7366	2.1563	0.07583	0.57583	1.2417	7.5937	13.187	16.375	5
6	0.08779	1.8244	2.5953	0.04812	0.54812	1.4225	11.390	20.781	29.562	6
7	0.05853	1.8829	2.9465	0.03108	0.53108	1.5648	17.085	32.171	50.343	7
8	0.03902	1.9219	3.2196	0.02030	0.52030	1.6751	25.628	49.257	82.515	8
9	0.02601	1.9479	3.4277	0.01335	0.51335	1.7596	38.443	74.886	131.77	9
10	0.01734	1.9653	3.5838	0.00882	0.50882	1.8235	57.665	113.33	206.66	10
11	0.01156	1.9768	3.6994	0.00585	0.50585	1.8713	86.497	170.99	319.99	11
12	0.00771	1.9845	3.7841	0.00388	0.50388	1.9067	129.74	257.49	490.98	12
13	0.00514	1.9897	3.8458	0.00258	0.50258	1.9328	194.62	387.23	748.47	13
14	0.00343	1.9931	3.8903	0.00172	0.50172	1.9518	291.92	581.85	1135.7	14
15	0.00228	1.9954	3.9223	0.00114	0.50114	1.9656	437.89	873.78	1717.5	15
16	0.00152	1.9969	3.9451	0.00076	0.50076	1.9756	656.84	1311.6	2591.3	16
17	0.00101	1.9979	3.9614	0.00051	0.50051	1.9827	985.26	1968.5	3903.0	17
18	0.00068	1.9986	3.9729	0.00034	0.50034	1.9878	1477.8	2953.7	5871.5	18
19	0.00045	1.9991	3.9810	0.00023	0.50023	1.9914	2216.8	4431.6	8825.3	19
20	0.00030	1.9994	3.9867	0.00015	0.50015	1.9939	3325.2	6648.5	13257.	20
21	0.00020	1.9996	3.9907	0.00010	0.50010	1.9957	4987.8	9973.7	19905.	21
22	0.00013	1.9997	3.9935	0.00007	0.50007	1.9970	7481.8	14961.	29879.	22
23	0.00009	1.9998	3.9955	0.00004	0.50004	1.9979	11222.	22443.	44841.	23
24	0.00006	1.9998	3.9969	0.00003	0.50003	1.9985	16834.	33666.	67284.	24
25	0.00004	1.9999	3.9978	0.00002	0.50002	1.9990	25251.	50500.	100951.	25
INF	0.00000	2.0000	4.0000	0.0000	0.50000	2.0000	INF	INF	INF	INF

Index